T0224807

RISING WATERS

The Causes and Consequences of Flooding in the United States

Hurricane Katrina and the flooding of New Orleans in 2005 confirmed what had been tacitly understood by local policy makers for a long time: that the problem of flooding is getting worse and that this presents a major risk to the health and safety of the American population.

This interdisciplinary book brings together over 5 years of empirical research funded by the National Science Foundation to explore the causes of flooding in the United States and the ways in which local communities can reduce the associated human casualties and property damage. Focusing on the two most vulnerable states in the nation, Texas and Florida, this book investigates factors other than rainfall that determine the degree of flooding, and considers the key role of non-structural techniques and strategies in flood mitigation. The authors present an empirical and multiscale assessment that underlines the critical importance of local planning and development decisions.

Written for advanced students and researchers in hazard mitigation, hydrology, geography, and environmental planning and public policy, this book will also provide policy makers, federal, state, and local government employees, and engineers with important insights into how to make their communities more resilient to the adverse impacts of chronic flooding in the future.

SAMUEL D. BRODY is a professor at Texas A&M University in the departments of marine sciences at Galveston and landscape architecture and urban planning at College Station. He holds the George P. Mitchell '40 Chair in Sustainable Coasts and is the Director of the Center for Texas Beaches and Shores. Dr Brody is also a faculty fellow at the Rice University Severe Storm Prediction Education and Evacuation from Disasters Center, the Texas A&M University Hazard Reduction and Recovery Center, and the Texas A&M University Institute for Science, Technology, and Public Policy. His research and teaching focus on coastal environmental planning, environmental dispute resolution, climate change policy, and natural hazards mitigation.

WESLEY E. HIGHFIELD is a research scientist at Texas A&M University at Galveston in the department of marine sciences. He is also the Associate Director of the Center for Texas Beaches and Shores. Dr. Highfield conducts research on the effects of wetland alteration on flooding and teaches courses in geographic information systems and biostatistics.

JUNG EUN KANG is a research fellow in the Korea Adaptation Center for Climate Change of the Korea Environment Institute. Dr Kang was formerly a postdoctoral research associate at the Hazard Reduction and Recovery Center at Texas A&M University where she conducted research on coastal flooding and natural hazard mitigation in the Gulf of Mexico.

RISING WATERS

The Causes and Consequences of Flooding in the United States

SAMUEL D. BRODY

Texas A&M University at Galveston and College Station

WESLEY E. HIGHFIELD

Texas A&M University at Galveston

JUNG EUN KANG

*Korea Adaptation Center for Climate Change
of the Korea Environment Institute*

CAMBRIDGE
UNIVERSITY PRESS

University Printing House, Cambridge CB2 8BS, United Kingdom

One Liberty Plaza, 20th Floor, New York, NY 10006, USA

477 Williamstown Road, Port Melbourne, VIC 3207, Australia

314-321, 3rd Floor, Plot 3, Splendor Forum, Jasola District Centre, New Delhi-110025, India

79 Anson Road, #06-04/06, Singapore 079906

Cambridge University Press is part of the University of Cambridge.

It furthers the University's mission by disseminating knowledge in the pursuit of education, learning and research at the highest international levels of excellence.

www.cambridge.org
Information on this title: www.cambridge.org/9781108446839

© Samuel D. Brody, Wesley E. Highfield, and Jung Eun Kang 2011

This publication is in copyright. Subject to statutory exception and to the provisions of relevant collective licensing agreements, no reproduction of any part may take place without the written permission of Cambridge University Press.

First published 2011
First paperback edition 2018

A catalogue record for this publication is available from the British Library

Library of Congress Cataloging in Publication data
Brody, Samuel David.
Rising waters : the causes and consequences of flooding in the United States /
Samuel D. Brody, Wesley E. Highfield, Jung Eun Kang.
p.   cm.
Includes bibliographical references.
ISBN 978-0-521-19321-4 (hardback)
1. Flood control–Government policy–United States.   2. Flood damage–United States.
3. Floods–United States.   I. Highfield, Wesley E.   II. Kang, Jung Eun.   III. Title.
TC423.B68 2011
363.34′930973–dc22
2010052186

ISBN   978-0-521-19321-4   Hardback
ISBN   978-1-108-44683-9   Paperback

Cambridge University Press has no responsibility for the persistence or accuracy of URLs for external or third-party internet websites referred to in this publication, and does not guarantee that any content on such websites is, or will remain, accurate or appropriate.

Contents

Preface

The information contained within this book was originally based on a National Science Foundation CAREER award to study the relationship between wetland alteration and coastal watershed flooding. Over the past six years, as dozens of colleagues, graduate and undergraduate students, and postdocs participated in the research, the scope of the project broadened to take on a more comprehensive look at flooding, flood impacts, and their policy implications. By bringing together multiple datasets and analyses, we aim to provide the reader with an evidence-based understanding of the degree to which flooding affects people's lives and the opportunities available to mitigate flooding impacts. To that end, we combine concepts, case studies, and statistical models to construct an overall vision of how to mitigate flood problems in the future. We focus our empirical study on Texas and Florida, but hope the findings and lessons learned throughout these pages can inform those based in other places.

While this book is scholarly in nature, it contains information useful to a broad audience, from those working in federal, state, and local government offices, to residents who have experienced flooding in their communities. Readers who do not wish to be overly burdened with understanding statistical procedures and relationships may skip to the summary sections at the end of each chapter, or to the policy recommendations presented at the conclusion of the book. Most of all, we hope that the material we present offers insight into the causes, consequences, and policy implications associated with floods in the United States.

Acknowledgments

This book is based on research supported in part by the U.S. National Science Foundation Grant No. CMS-0346673 to Texas A&M University, the U.S. Environmental Protection Agency STAR Grant No. FP-91661001, and the Houston Endowment. The findings and opinions reported are those of the authors and are not necessarily endorsed by the funding organizations or those who provided assistance with various aspects of the study.

We would like to thank the many research assistants, postdocs, and faculty who have contributed to various aspects of the research contained in this book. These include: Himanshu Grover, Hyung-Cheal Ru, Sarah Bernhardt, Laura Spanel-Weber, Sammy Zahran, Praveen Maghelal, Zhenghong Tang, Anita Hollman, Beth Larkin, Amanda Jaloway, Walter M. Peacock, Scott Burleigh, Linda Salzar, and Ching-Yu Chou. Without their time and input this research could have never been accomplished. We are all indebted to Jennifer Feagins for her laudable organizational skills, attention to detail, and ability to bring this work together in its final form.

We also thank our colleagues and mentors for their encouragement and advice. Appreciation goes to Walter G. Peacock, Michael Lindell, Phil Berke, Dennis Wenger, Phil Bedient, Shannon Van Zandt, Arnold Vedlitz, Bill Merrell, and Bill Seitz. Special thanks go to our editors Ruth Schemmer and Linda Brody who made the forthcoming pages so much more readable.

Finally, and most importantly, we want to thank our friends and family for their constant patience, love, and support for all of our scholarly endeavors.

Abbreviations

ACE	Army Corps of Engineers
ASFPM	Association of State Floodplain Managers
CBD	central business district
CBO	Congressional Budget Office
CRS	community rating system
CSTS	cross-sectional time-series
CWA	Clean Water Act
DCA	Department of Community Affairs
(the) District	Harris County Flood Control District
EPA	Environmental Protection Agency
FEMA	Federal Emergency Management Agency
FIRM	flood insurance rate map
FMA	Flood Mitigation Assistance (Program)
GAO	Government Accountability Office
GIS	geographic information systems
HGMP	Hazard Grant Mitigation Program
HUD	Housing and Urban Development
ICC	increased cost of compliance
NCDC	National Climatic Data Center
NESDIS	National Environmental Data and Information Service
NFIP	National Flood Insurance Program
NOAA	National Oceanic and Atmospheric Administration
NRC	National Research Council
NSF	National Science Foundation
NWS	National Weather Service
PDM	Pre-Disaster Mitigation (Program)
RFC	Repetitive Flood Claims (Program)

SHELDUS	Spatial Hazard Events and Losses Database for the United States
SFHA	special flood hazard area
SRL	Severe Repetitive Loss (Program)
USACE	U.S. Army Corps of Engineers
USDA	U.S. Department of Agriculture
USGS	U.S. Geological Service

1

Introduction: rising waters

On June 20, 2006, a heavy summer storm dumped over 10 inches (25.4 cm) of rain on the Houston, Texas, metropolitan area: several bayous overflowed and there were over 500 emergency calls for help from motorists stranded on flooded roads. On October 16 of the same year, only 6.5 inches (16.5 cm) of rainfall was enough to submerge a highway underpass leading to a major interstate under almost 12 ft (3.66 m) of water. As a consequence of the storm, schools closed and four people died when they became trapped in their cars. Three years later, in April 2009, the same amount of precipitation fell across Houston, once again flooding homes and major roads. This time, approximately 200 motorists were marooned in a parking lot, 80% of streets were under water in several neighborhoods, and five children were among those who died. This chronic pattern of flooding, property damage, and human casualties is not unique to Houston, but is replicated in thousands of towns and cities across the U.S. Rapid population growth in low-lying coastal areas, sprawling development patterns, and the alteration of hydrological systems are just three of the factors shaping the development of flood-prone communities in which more people are being placed at risk.

The little-known fact is that, among all natural hazards, floods pose the greatest threat to the property, safety, and economic well-being of communities in the U.S. More property is lost and more people die from flood events than from tornados, earthquakes, and wildfires combined. And, despite federal policies created to guide both structural and non-structural mitigation initiatives, property damage and human casualties continue to mount across the nation. Apart from the sheer disruption caused to people's lives, the economic impact alone is estimated in billions of dollars annually (Association of State Floodplain Managers [ASFPM], 2000; Pielke, 1996). According to data extracted from the Spatial Hazard Events and Losses Database for the United States (SHELDUS), damage from floods has increased over time. In the 1960s, flood damage averaged $45.65 million a year; by the 1990s, average annual property damage from flooding increased to

1

$19.13 billion dollars a year (inflation adjusted at 1960 dollars (Brody *et al.*, 2007a: 330–345). One reason for the tremendous increase in property damage may be the occurrence of floods: according to SHELDUS, the average annual flood count has increased sixfold, from 394 floods per year in the 1960s to 2444 flood events per year in the 1990s.

It is not necessary to conduct a national study of flood events and their impacts to elucidate what local decision makers and residents have tacitly understood for decades: Floods pose a major risk to the health and safety of U.S. communities and this problem is only becoming worse. Perhaps the real threat comes not from major storms, but from the multitude of chronic, small-scale floods that barely make headlines. Catastrophic events such as the great floods of the Midwest in 1993 and Tropical Storm Allison in 2001 are well documented and quickly addressed. But the relatively small storms that come and go almost undetected by the media and the public add up to billions of dollars in losses over time. A single event, such as the 2009 late afternoon thunderstorm described above, which passed over Houston, is not by itself a cause for alarm. Rather, it is the cumulative impact of these repetitive, small-scale events that weigh down local economies, tear at the fabric of community well-being, and disrupt the daily lives of millions of residents.

In *Disasters by Design*, Mileti (1999) argues that disasters do not simply occur as acts of God, but are instead largely the result of how we build and design human communities. This notion has helped many scholars and decision makers realize that disasters are literally human-constructed events that can be mitigated through thoughtful land use policies. However, this concept has gone largely untested from an empirical perspective among researchers over the past decade.

This book builds on Mileti's theory by offering systematic, empirical evidence that the location, intensity, and pattern of the built environment are critical factors in determining the impacts of floods. Our underlying premise is that the rising cost of floods is not solely a consequence of increasing mean annual precipitation, population growth, or inflationary monetary systems. It is also driven by the manner in which we plan for and subsequently develop the physical landscape. Individual and community-based decisions pertaining to the distribution of buildings and impervious surfaces, and the degree to which hydrological systems are altered, are exacerbating losses from repetitive floods. Increasing development associated with residential, commercial, and tourism activities, particularly in coastal and low-lying areas, has diminished the capacity of hydrological systems (e.g., watersheds) to naturally absorb, hold, and slowly release surface water runoff. As a result, private property, households, businesses, and the overall economic well-being of coastal communities have become increasingly vulnerable to the risks of repetitive flooding events.

In the U.S., decisions about planning and development reside at the local level, and therefore effective flood mitigation lies in the hands of county commissions, zoning boards, mayors, planning departments, and other local governmental entities. From this perspective, flood control and avoidance can no longer be considered the sole province of the federal government. The twenty-first-century vehicles for preventing loss of property and life may not be based only on federal disaster relief or large dams and levees, but also on county and citywide land use plans, development and construction codes, zoning and subdivision ordinances, technical assistance, community-based outreach, and other locally based non-structural programs.

Our book rigorously examines the causes, consequences and policy implications of repetitive flooding in the U.S. This book is the culmination of over five years of empirical research funded by the National Science Foundation on how local communities can reduce both property damage and human casualties associated with flooding. Focusing on two of the most vulnerable states in the nation, Texas and Florida, we have fully investigated the factors contributing to the degree of flooding and how local planning and development decisions may be critical elements in determining the extent of damage experienced by local communities. It is our proposition that, over time, local communities in the U.S. have increasingly borne the responsibility for flood problems. By adopting and implementing both structural and non-structural mitigation measures, localities have taken important steps to reduce property damage and human casualties associated with localized flood events. With this understanding, we examine the effects of past decisions on floods and flood damage, in order to offer future solutions that could more effectively mitigate the adverse impacts of flood events.

Research questions

Recognizing that now, more than ever, scientific knowledge is needed to better comprehend the impacts of development-based decisions on flooding, we posit that, with better knowledge, local decision makers can reduce loss of both property and human lives in the future. While much research has catalogued the amount of damage caused by floods in the U.S., there is comparatively little work on the local-level causes, consequences, and policy implications associated with repetitive flooding events. The inquiry that exists is largely argumentative, focused on a single time period, and based on isolated case studies. This research approach makes it difficult to externalize findings that may be useful to a broader community of policy makers. In fact, over the past decade, no systematic, large-scale, quantitative study has been conducted that can move the field of flood mitigation and planning in the U.S. forward.

This book directly addresses this research need by providing empirically driven, evidence-based results that can directly inform local decision makers and scholars about effectively reducing the adverse impacts of floods. We employ multiple methods of inquiry, including geographic information systems (GIS) analytical techniques, multivariate modeling, and surveys to thoroughly investigate the characteristics, causes, consequences, and policy implications of flooding throughout coastal Texas and Florida. The primary research questions addressed in the following chapters are:

- What is the spatial pattern of local flooding in coastal Texas and Florida?
- What is the effect of development on the degree of flooding and flood damage?
- To what degree does the alteration of naturally occurring wetlands along the coast contribute to riparian flooding and property damage from floods?
- How well are coastal communities preparing themselves to mitigate repetitive flooding events?
- To what degree are local flood mitigation and planning working in terms of saving lives and property?
- How does local organizational capacity influence the level of preparedness and mitigation to reduce the adverse impacts of floods?
- What are the major factors driving households to insure themselves against flood risks?

Study areas: why Texas and Florida

For the study areas, we selected coastal Texas and Florida (Figure 1.1) to examine the characteristics, causes, and consequences of flooding and flood mitigation for several reasons. First, both states are situated within the coastal zone of the Gulf of Mexico, where population growth and development make communities vulnerable to the effects of flooding. Given the recreational, aesthetic, and economic opportunities available on the coast, this geographic area has historically been the focus for extensive population growth and land use change. In 2003, for example, it was estimated that approximately 153 million people (53% of the U.S. population) live in the 673 coastal counties, an increase of 33 million people since 1980 (Crossett *et al.*, 2004). With increasing population come more structures located in areas susceptible to flooding from severe storms that routinely strike coastal areas. From 1999 to 2003, 2.8 million building permits were issued for the construction of single-family housing units (43% of the nation's total) and 1 million building permits were issued for the construction of multi-family housing units (51% of the nation's total) within coastal counties across the U.S. (Crossett *et al.*, 2004). Because communities positioned along a coastline or within a coastal watershed are especially

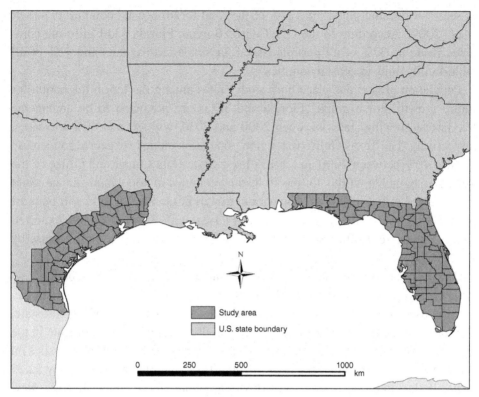

Figure 1.1 The study areas.

vulnerable to flooding from both severe and chronic storm events, this upward trajectory of growth has created the ideal conditions for human catastrophe.

In Texas, between 1980 and 2003, approximately 2.5 million people more moved into the coastal areas, representing a 53% increase. Only California and Florida ranked higher in the addition of coastal population (Crossett *et al.*, 2004). Surprisingly, from a total land area perspective, the Texas coast remains relatively undeveloped, with the population centered in the Houston–Galveston area, Corpus Christi, Beaumont, and Brownsville. However, these populated areas have recently become places of intense growth. For example, from 1980 to 2003, Harris County, where Houston is located, had the second highest coastal growth rate among all counties in the U.S. (Crossett *et al.*, 2004). Today, the Houston–Galveston–Brazoria region remains one of the fastest growing areas in the country, with over 2100 persons per square mile.

Florida has historically been a place of intense population growth and development, and its pace continues to rank among the top in the nation. For example, in 2000, Florida, along with California and New York, comprised 41% of the total housing units added among all coastal counties. Florida also had the largest number of seasonal housing units, over 24% of the total for all coastal counties (Crossett *et al.*, 2004). According to the U.S. Census Bureau, Florida and California combined made up 37% of all permits issued for single-family units and 42% of all multi-family units in coastal counties.

Population projections place both study states among the top in the nation for future growth. For example, Florida and Texas are predicted to be among the five fastest-growing states between 2000 and 2030, with 80 and 60% increases, respectively. The Texas Gulf coast region population alone is expected to increase by over 40% between 2000 and 2015 (Texas State Data Center and Office of the State Demographer, 2008). Of the 10 leading counties in population change, eight are expected to be in Florida. Population growth in the next decade will be most prominent in the southernmost portion of Florida, with Broward County expected to increase by 167 000 persons and Palm Beach County expected to increase by 151 000.

A second rationale for selecting our study areas is that both states contain low-lying coastal areas, which makes them extremely prone to flooding. For example, Texas consistently incurs more deaths (double the total for the second-highest state, California) and insurance losses per year from flooding than any other state in the U.S. According to Federal Emergency Management Agency (FEMA) statistics on flood insurance payments from 1978 to 2001, Texas reported approximately $2.25 billion dollars in property loss. These losses amount to more than California, New York, and Florida combined, the next three states on the list of the most property damage (National Flood Insurance Program [NFIP], 2007). Florida also experiences significant annual economic losses from floods due to its low elevation, large coastal population, and frequent storms. For example, based on a composite flood risk measure combining floodplain area, population, and household values, Florida ranks the highest among all states (FEMA, 1997). Recent estimates indicate that from 1990 to 2003, the state suffered losses of almost $2.5 billion (in current U.S. dollars).

A third reason for selecting Texas and Florida as our study areas is that while both states are similarly susceptible to flooding and flood damage, their different policy settings and development patterns allow for an insightful comparative analysis. For example, with the passage of the Florida Growth Management Act in 1985, Florida adopted a statewide mandate requiring all local jurisdictions to adopt a legally binding, prescriptive comprehensive land use plan (Chapin *et al.*, 2007). Under this requirement, cities and counties within the state must adopt in their plans flood mitigation and coastal natural hazard policies. Florida's top-down mandate and regulatory environment for development has long been considered one of the strongest in the nation. Rule 9J-5, adopted by the Department of Community Affairs (DCA) in 1986, requires that specific elements and goals be included in local plans and prescribes methods local governments must use in preparing and submitting plans. Despite this "checklist" approach to land use planning, there is evidence of wide disparities in the breadth and quality of local environmental policies in Florida (see Brody, 2003a, for more detail).

In contrast, Texas has no comparable state-level planning mandate. Land use planning, development regulations, and flood mitigation are largely the responsibility of local jurisdictions. As a result, flood planning and mitigation activities are spotty along the Texas coast. A Texas culture of strong private property rights combined with a lack of government regulations concerning development patterns versus the centralized Florida model provides an ideal comparative setting in which to examine the effectiveness of flood plans and policies.

Outline of the book

The book is categorized into four related parts. Part I examines the consequences of flooding in the U.S. Chapter 2 addresses trends in flood damage and casualties from a national perspective and at multiple spatial scales. We cover past studies on the status of flooding, and provide new data enabling us to make more current and spatially specific damage estimates. Chapter 3 focuses on local communities within the study area states of Texas and Florida. Here, we provide more detailed information regarding property damage, injuries and fatalities, insurance purchase amounts, physical risk variables, etc. In this chapter, we focus the reader's attention on specific geographic areas, which serve as the representative target for the rest of the book. After examining the trends and status of flood impacts across the country and within the study area, Chapter 4 critically examines the existing legal and policy frameworks associated with flood mitigation. Special attention is paid to the role of the NFIP, FEMA's community rating system (CRS), and specific local mitigation initiatives.

Part II of the book concentrates on the major factors influencing the amount of flooding, property damage, and human casualties from flood events. Chapter 5 lays the groundwork for explaining the causes and consequences of flood impacts, and the policy implications of hazard mitigation at the local level. Specifically, we identify and discuss the major factors influencing flooding and flood damage, including the natural environment, socioeconomic factors, the built environment, and mitigation. In Chapter 6, we focus on the degree to which naturally occurring wetlands mitigate flooding and the human impacts of the floods. First, we trace the policy and regulatory history behind wetland development in the U.S. Second, we discuss the current permitting process administered by the Army Corps of Engineers (ACE), which includes different types of permits suited for specific development conditions. Third, we present the results of our own inquiry on the spatial pattern of wetland development over a 13-year period, the different types of wetland altered along coastal Texas and Florida, and most importantly the contribution of wetland permit issuance to flooding and flood damage across the study area. Chapter 7 focuses on the role of FEMA's CRS as a proxy for non-structural local flood mitigation activities. After providing an overview of the program, we describe the extent to which local policies are adopted within the study areas and explain the relationship between policy implementation and reduced property damage from floods. We also report the results of a survey of planners and managers conducted in both states on the effectiveness of specific policies implemented at the local level. In Chapter 8, we examine the effects of contextual community and development factors on flooding. Empirical analyses identify the effects of structural mitigation, such as dams and levees; the role of impervious surface resulting from increased urban development; socioeconomic variables, such as community wealth and education; and demographic variables, including population, population growth, and housing density.

Part III of the book focuses on the role of learning in terms of reducing the adverse impacts of floods within our study area. By taking a longitudinal approach, we unravel policy learning and adjustment at the community, institutional, and household levels. Chapter 9 investigates the drivers of policy change instituted by governments, as well as household adjustments associated with insurance purchasing. Quantitative models demonstrate the degree to which communities are improving their mitigation capabilities in the face of continual flood events and the reasons localities and households are willing to make these changes. Because policy change and learning, in response to repetitive flood events, comprise such a complex issue based on contextual characteristics, in Chapter 10 we supplement quantitative findings with case studies. Several cases based on secondary documentation and interviews with planning officials from jurisdictions within the study areas add depth to the arguments based on statistical analysis made in the previous chapter.

Finally, Part IV of the book discusses the policy implications of the research findings and presents a set of integrated planning recommendations for improving the ability of local communities to reduce the negative effects of flooding across the U.S. In both Chapters 11 and 12, we extend our results for Texas and Florida to coastal communities in general and set forth a policy agenda for flood mitigation in the twenty-first century, focusing on local level initiatives. We also identify specific future research needs that can help move the U.S. closer to building more resilient communities in the future.

Part I

The consequences of floods

2

Rising cost of floods in the United States

The U.S. is assaulted by a variety of natural hazards every year, totaling tens of billions of dollars in direct damages. Hurricanes along the coasts, earthquakes in California, blizzards in the Midwest, tornadoes, wildfires, and drought are just some of the hazards contributing to persistent, economically damaging adverse impacts. What is virtually unknown and rarely discussed in the public domain is that among all natural hazards, floods pose the greatest threat to the property, safety, and economic well-being of local communities across the nation. While these events rarely make the national news for their drama and intensity, in total, economic impacts from floods alone are estimated in billions of dollars each year (ASFPM, 2000; Pielke, 1996).

Floods have always plagued metropolitan areas in the U.S. because cities tend to be located along major river bodies and in coastal areas for better access to commerce. But, a casual glance at any flood dataset shows that the problem is getting worse, despite the fact that Americans have steadily moved out of urban centers into sprawling suburban environments. For example, data from SHELDUS show that the number of floods per year has increased sixfold, from an average of 394 floods per year in the 1960s to 2444 in the 1990s. This dataset also reveals that floods in the 1960s caused approximately $41 million of property damage a year compared with over $378 million dollars a year in the 1990s (inflation adjusted to 1960 dollars).

This chapter lays the foundation for the remainder of the book by examining direct damages attributed to flooding in the U.S. using multiple sources of data. Specifically, we investigate the magnitude and trends for human causalities, crop damage, and property damage caused by floods at both national and state levels. We report on past flood studies and provide new data, enabling us to paint a more current and spatially specific picture of the consequences of floods nationwide. Our objective is not to rewrite or replace previous studies calculating the impacts of floods, but rather provide a descriptive basis to support the content of the remaining chapters, which discuss how to respond to the increasing threat of floods.

Understanding data on flood losses

Floods cause a range of adverse impacts, including human injuries and fatalities, and damages to crop, property, and public infrastructure. Losses involve both direct and indirect costs. Direct costs are related to immediate physical damage and repair expenditures generated by flooding events. Indirect costs include loss of business and personal income, reduction in property value, reductions in tax revenue, psychological impacts, and depletion of ecosystem services (Heinz Center, 2000). Comprehensive assessments of the impacts from floods should include both direct and indirect measures. However, because indirect loss is very difficult to identify and measure at a national level, currently available flood databases in the U.S. focus almost exclusively on direct costs, such as property and crop damage, and the number of human casualties caused by an event. We take advantage of existing data streams by focusing on direct costs, but it should be noted that the actual impact from flooding in the U.S. is most likely more extensive than reported in this chapter.

Currently, there are four major flood loss databases: National Weather Service (NWS); reanalysis of NWS damage; SHELDUS; and NFIP. To better understand the nature of flood loss in the U.S., we review and compare each dataset. Table 2.1 summarizes the four data sources showing variation in collection techniques, accuracy, accessible timelines, and spatial scales.

The NWS collects flood loss data through field offices; thus, the quality of loss estimates may vary depending on the procedures followed in each particular office. The goal of this agency is not to assign entirely accurate damage figures, but to predict flood events which lead to losses (NWS, 2004). That being said, the NWS provides fatalities, injuries, property damage, and crop damage estimates at national and state levels from 1995 to the present on the Natural Hazard Statistics website (www.weather.gov/os/hazstats.shtml). NWS loss estimates include only direct damage due to flooding caused by rainfall and/or snowmelt. The dataset excludes flooding generated by winds from hurricanes, storm surges, tsunami activity, and coastal flooding because "although they cause water inundation, they are not hydro-meteorological events" (Pielke *et al.*, 2002).

Pielke *et al.* reanalyzed the NWS flood damage data in 2002 through a project sponsored by the National Science Foundation (NSF) and the National Oceanic and Atmospheric Administration (NOAA). The purpose of this project was to correct previous inconsistencies and derive a more complete and dependable assessment of flood damages in the U.S. These data (www.flooddamagedata.org/national.html) are available at national, state, and watershed/basin levels from 1926 to 2003.

Table 2.1 *Data sources of flood losses and human casualties*

Source	Spatial scale	Available time period (by November 2009)	Provided data	Author/available information
National Weather Service flood damage dataset	National State	1995–2008	Fatalities, injuries, property damage, crop damage	National Weather Service, Office of Climate, Water and Weather Services
Reanalysis of National Weather Service flood damage dataset	National State Basin	1926–2003	Property loss	Pielke *et al.*, (2002)
Spatial Hazard Events and Losses Database for the United States dataset	National State County	1960–2008	Property and crop losses, injuries and fatalities.	Hazard Research Lab at the University of South Carolina Database of 18 natural hazards including floods Date, location
National Flood Insurance Program dataset	National State Jurisdiction (Community)	1969–2007 (1978–2007 data available on the internet)	Insured property damage	Federal Emergency Management Agency Loss dollars paid, number of claims paid, policies in force, premium and coverage

SHELDUS is another source of flood impact data housed in the Hazard Research Laboratory at the University of South Carolina. This database (www.sheldus.org) reports county-level damages associated with floods and 17 other natural hazards from 1995 to 2008. Flood losses are derived from monthly storm data publications prepared by the National Climatic Data Center (NCDC), the National Environmental Data and Information Service (NESDIS), NOAA, and the NWS (Hazards Research Lab, 2008). The "Storm Data" published by the NCDC assesses flash flood and riverine flood events, which are associated with heavy rainfall and/or snowmelt. SHELDUS data separate rainfall-based flooding from floods caused by storm surge or high tides. This database is useful in that detailed information on the day of each flood event is provided at the county level.

The final dataset we examined is from the NFIP, established in 1968 to provide flood insurance to residents in participating localities (see Chapter 4 for more details). The program was originally adopted to provide insurance against flood losses as an alternative to federal relief (Pasterick, 1998: 125–155). Losses paid by the NFIP cover inundation from tidal water as well as rainfall. For our purposes the record of paid claims for flood losses provides a very good measure of the amount of damage incurred within communities, particularly for residential properties. Because these data are composed of precise figures rather than broad estimates, they offer valuable tools for understanding the impacts of floods and making policy recommendations to reduce future losses. Using these data, we can also assess the impact of floods from both rainfall and tidal events, which other sources segregate.

The greatest shortcoming of the NFIP data stream, however, is it represents only insured losses. In addition, the presence of government-based flood insurance for an individual requires the community to voluntarily participate in the NFIP. Furthermore, the community must enforce NFIP mitigation requirements to remain in the program. In the beginning stages of the NFIP, participation rates were very low, but they have steadily increased, particularly after 1973, when the Flood Disaster Protection Act strengthened the NFIP by mandating that community participation in NFIP was a condition for being eligible to receive certain types of federal assistance. By 1977, approximately 15 000 communities joined the NFIP (FEMA, 2002). As of 2007, nearly 20 000 communities across the U.S. and its territories were voluntarily participating in this program (FEMA, 2007a). In spite of increased participation in the program, NFIP data still capture only insured losses, as opposed to actual losses.

Each data source described above catalogues flood losses using different methods, timespans, and spatial scales. The datasets also have their own set of biases and inconsistencies, making direct comparisons problematic (Gall *et al.*, 2009: 799–809). As shown in Figure 2.1, national losses associated with property damage (in actual dollars between 1960 and 2007) show different yearly intensities over time. Annual flood losses fluctuate, but the historical trend from the 1960s to 2000s indicates an increase in damages of between $86 million and $16 billion per year. Given the discrepancies among the four databases, stemming from different methodologies of data collection, definition of flood loss, time periods, and overall reliability, it is difficult to triangulate an overall trend of flood damages. However, it is apparent that flood damage has increased over the long term with a growing number of catastrophic flooding events, such as the Midwest flood in 1993 and Hurricane Katrina in 2005. When strong floods occur

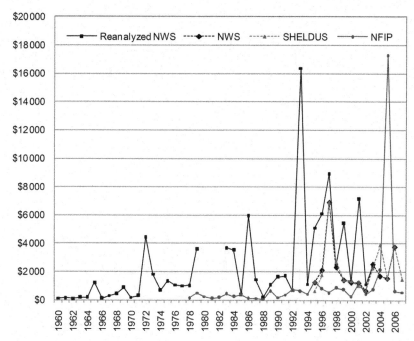

Figure 2.1 Trends of flood damage to property in the U.S.

in highly populated and unprepared communities, they tend to result in huge financial losses.

Here, we emphasize some researchers' (Mileti, 1999; Pielke *et al.*, 2002) arguments that the techniques and models currently at hand are insufficient; additional and more reliable flood-related loss estimation is a necessity. These authors have also proposed the need for a single agency whose responsibility is to collect, manage, and release official flood damage datasets. Accurate data at multiple spatial scales will greatly enhance our ability to assess the status and trends associated with flooding, which will support more effective decision making over the long term.

Human casualties from floods

The most severe impact from floods is human death and injury. In general, very few people are aware of just how many human casualties result from flooding events. According to the SHELDUS data, between 1960 and 2008, 3972 people lost their lives and over 17 751 people were injured from floods. As shown in Figure 2.2, the frequency of flood fatalities per year varies widely, but has actually been decreasing nationwide since the 1970s. For example, the average annual fatality rate during the 2000s was approximately 70 people compared to 118 in

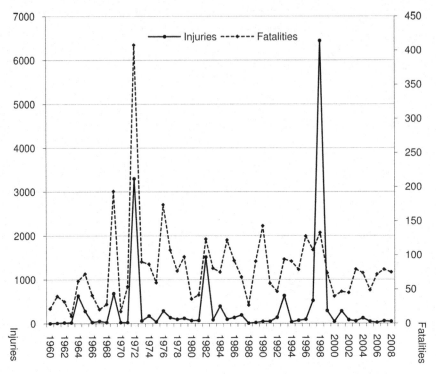

Figure 2.2 Trends of flood fatalities and injuries, 1960–2008.

the 1970s. A spike in flood fatalities occurred in 1972 (increasing the national average to 408 people) when a dam failure near Rapid City, South Dakota, killed 237 people (Ashley and Ashley, 2008: 805–818). This accident was reported as the most deadly single flood event in terms of casualties, until Hurricane Katrina in 2005. Figure 2.2 also depicts the national trend for flood injuries from 1960 to 2008. Similar to fatalities, there are spikes in the number of injuries based on iso-lated flood events. The trend line shows three anomalously high years. In 1998, of 6446 injuries caused by floods, 98.6% occurred in Texas when the Guadalupe River overflowed its banks after heavy rainfall. The flooding in South Dakota mentioned above also caused 2932 injuries in 1972. Finally, in 1982, California experienced 1081 injuries (71.45 of the total injuries in the U.S. for that year) from persistent riverine flooding.

Examining the spatial distribution of flood casualties indicates which states are most vulnerable to both human fatalities and injuries over the 48-year time period. As shown in Figure 2.3 and Table 2.2, Texas incurred by far the most

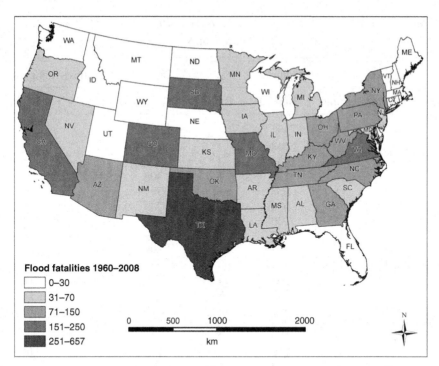

Figure 2.3 Flood fatalities by state, 1960–2008.

fatalities (657), followed by South Dakota (246), Virginia (224), California (186), and Colorado (185). In contrast, the safest states with regard to floods were Rhode Island, Idaho, and the District of Columbia. Texas also led the nation in injuries (Figure 2.4) caused by floods (7440), well ahead of South Dakota (3194), California (1576), Iowa (808), and Kentucky (797).

To demonstrate how much of an outlier Texas is for harmful floods compared with the rest of the U.S., we combined fatalities and injuries to derive a casualties figure (Table 2.2). The 8000+ flood casualties observed in Texas from 1960 to 2008 are approximately 2.5 times those in the second state on the list (South Dakota) and almost 332 times the casualty sum of the five lowest states. Why are so many people hurt or killed by floods in Texas compared with any other state? It cannot be based solely on population, precipitation, or area of floodplain alone. From a physical vulnerability standpoint, Florida tops the list, but ranks only 38th in human casualties among all states. This question of loss and vulnerability is something we will investigate in future chapters to better understand the factors driving flood damages within our study areas.

The consequences of floods

Table 2.2 *Flood fatalities, injuries, and casualties for the 51 U.S. states,*
1960 to 2008

State	Fatalities Number	Rank	Injuries Number	Rank	Casualties (fatalities + injuries) Number	Rank
TX	657	**1**	7441	**1**	8098	**1**
SD	246	**2**	2948	**2**	3194	**2**
CA	186	**4**	1390	**3**	1576	**3**
IA	36	31	772	**4**	808	**4**
KY	139	9	658	**5**	797	**5**
OH	147	8	638	6	785	6
CO	185	**5**	306	8	491	7
NY	117	13	312	7	429	8
VA	224	**3**	175	16	399	9
AZ	127	12	254	14	381	10
OK	109	14	263	11	372	11
MN	55	23	275	9	330	12
PA	150	7	178	15	328	13
LA	53	25	273	10	327	14
NJ	45	26	258	13	303	15
VT	16	37	259	12	275	16
MO	167	6	57	27	224	17
NC	104	15	108	17	212	18
TN	130	11	79	21	209	19
WV	133	10	64	26	197	20
GA	103	16	90	20	193	21
MD	55	24	106	18	161	22
AR	59	19	96	19	155	23
NM	65	18	78	22	143	24
IN	69	17	70	25	139	25
NV	37	30	73	24	110	26
AL	58	21	51	30	109	27
IL	58	20	49	31	107	28
WY	16	38	75	23	91	29
SC	38	29	51	29	89	30
UT	30	34	55	28	85	31
OR	56	22	24	35	80	32
MI	36	32	42	32	78	33
KS	40	28	34	33	74	34
MS	40	27	22	36	62	35
HI	34	33	22	37	56	36
WI	16	39	28	34	44	37
FL	23	36	8	41	31	38
WA	24	35	5	44	29	39
NE	13	41	14	39	27	40
NH	9	44	17	38	26	41
CT	15	40	5	42	20	42

Table 2.2 (*cont.*)

State	Fatalities		Injuries		Casualties (fatalities + injuries)	
	Number	Rank	Number	Rank	Number	Rank
DE	10	42	9	40	19	43
ND	10	43	5	43	15	44
MT	8	46	3	46	11	45
ME	7	47	4	45	11	46
MA	6	48	3	47	9	47
AK	9	45	0	50	9	48
DC	1	49	3	48	4	49
ID	1	50	2	49	3	50
RI	0	51	0	51	0	51
Total	3972		17 751		21 724	

Adapted from Hazards Research Lab: SHELDUS (http://webra.cas.sc.edu/hvri/products/sheldus.aspx; accessed February 1, 2010).

Crop damage and property damage

Despite efforts to mitigate the economic impacts of floods, losses have increased over time, even when adjusting for inflation. As previously mentioned, floods are the most costly natural hazard afflicting the U.S. In this section, we address the magnitude of this cost.

One of the most devastating and unrecognized impacts from floods is loss associated with agricultural operations. During the 48-year period from 1960 to 2008, the U.S. experienced over $11 billion damage to crops due to flooding. Annual crop damage fluctuates from year to year, but the historic trend from the 1960s to 2000s shows an overall increase (see Figure 2.5). The U.S. had unusually high crop damages of over $1.3 billion in 1993, 1995, and 2008. In 1993, for example, severe floods in the Midwest caused catastrophic damage to the farming community. According to the data, Iowa alone reported over $1 billion in crop damage and the damage in Illinois totaled almost $500 million. More recently, in 2008, Indiana sustained $600 million in crop damage, while Iowa and Wisconsin reported approximately $300 million of flood-related damage. From a spatial perspective (Figure 2.6), certain Midwestern states (Iowa, Illinois, Indiana, and Minnesota), along with Texas, Florida, and California, where agricultural activities are most prominent, were hotspots for crop damage caused by floods.

Flood-related property damage in the U.S. has also steadily increased over the past four decades (see Figure 2.5). In fact, average annual damage has increased

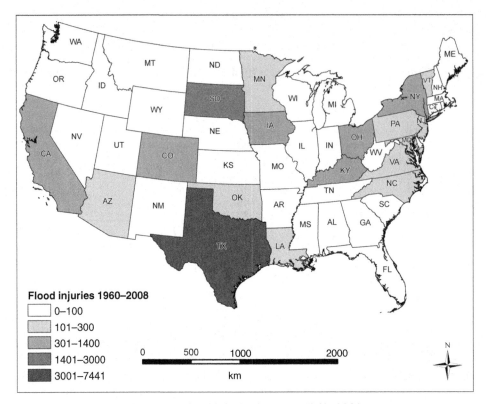

Figure 2.4 Flood injuries by state, 1960–2008.

54 times during this period, from $51 million in 1960s to $2.77 billion per year in the 2000s (2000–2008). According to the SHELDUS database, 2008 had the largest amount of property loss from floods, primarily because several loca-tions in the Midwest experienced multiple floods caused by prolonged and heavy precipitation that overflowed banks or breached levees. These floods affected Illinois, Indiana, Iowa, Michigan, Minnesota, Missouri, and Wisconsin. Another major year for flood losses was 1997, during which the nation suffered approxi-mately $7 billion in property damage. This year was characterized by the Red River flood, which occurred along the Red River in North Dakota and Minnesota (along with Southern Manitoba, Canada), resulting in about $3.5 billion in prop-erty damage.

Figure 2.7 illustrates the distribution of flood property damage by state, total-ing $52 billion (in constant dollars) from 1960 to 2008. The top five states for property damage are Iowa, North Dakota, Pennsylvania, Texas, and New York.

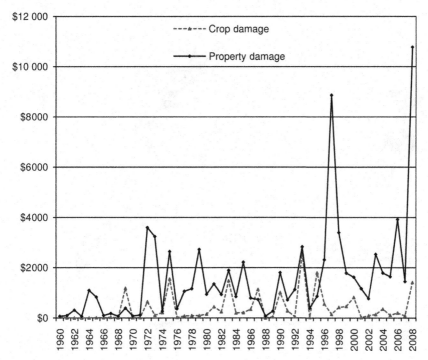

Figure 2.5 Trends in flood crop and property damage, 1960–2008.

Iowa suffered almost $9.4 billion in property loss from 1960 to 2008, which is over 2.5 times that for the second state on the list, North Dakota ($3.78 billion), and approximately 1202 times Rhode Island, the state with the lowest amount of damage. In general, the bulk of crop and property damage in the U.S. occurs in just a few states, likely for different reasons (Table 2.3). In fact, the top 10 states on the list represent almost 60% of total assessed damage from floods.

Generally, states with larger populations and more urbanized land will experience a larger amount of property damage. The top 10 states for property damage include four of the most highly populated states – California, Texas, Florida, and New York. However, population is clearly not the only factor explaining flood-related property losses. For example, both Iowa and North Dakota, which are also in the top 10 states for property damage, do not have large populations, but have major river systems that tend to flood in heavy rain. When we standardize crop and property damage by population in the year 2000, North Dakota and Iowa remain at the top of the rankings. However, Mississippi, West Virginia, Maine, and Vermont then also rank among the top 10 damaged states. Large amounts of

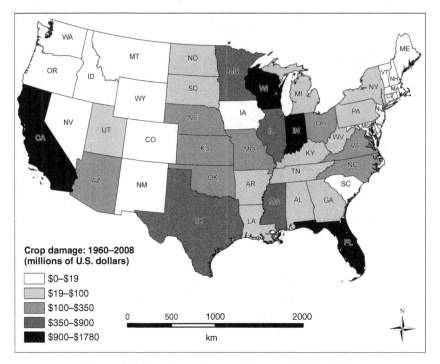

Figure 2.6 Flood crop damage by state, 1960–2008.

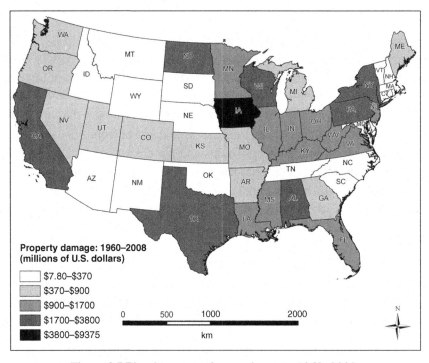

Figure 2.7 Flood property damage by state, 1960–2008.

Table 2.3 *Flood crop damage and property damage by U.S. state, 1960–2008*

State	Crop damage Amount	Rank	Property damage Amount	Rank	Crop and property damage Amount	Rank
IA	1 780 130 875.12	1	9 374 310 926.74	1	11 154 441 801.86	1
CA	1 439 387 000.05	2	2 433 667 571.35	7	3 873 054 571.40	2
ND	82 340 000.00	20	3 780 717 499.94	2	3 863 057 499.94	3
WI	1 329 878 917.78	3	2 480 769 226.39	6	3 810 648 144.17	4
TX	753 629 235.12	6	2 620 834 383.38	4	3 374 463 618.50	5
PA	53 812 923.32	26	2 623 415 751.33	3	2 677 228 674.65	6
NY	65 042 983.11	23	2 528 606 885.20	5	2 593 649 868.31	7
FL	933 945 992.10	4	1 477 959 894.45	13	2 411 905 886.55	8
IN	906 675 549.79	5	1 284 283 629.54	15	2 190 959 179.33	9
MN	525 934 300.02	9	1 642 741 547.60	9	2 168 675 847.62	10
MS	585 919 375.86	8	1 554 175 966.16	11	2 140 095 342.02	11
OH	332 549 559.16	10	1 582 225 276.09	10	1 914 774 835.25	12
AL	39 311 500.11	30	1 719 341 999.68	8	1 758 653 499.79	13
IL	589 524 239.31	7	1 145 752 573.16	18	1 735 276 812.47	14
LA	56 210 000.62	24	1 516 386 752.99	12	1 572 596 753.61	15
VA	195 047 884.59	13	1 229 714 147.26	17	1 424 762 031.85	16
NJ	800 384.60	44	1 415 663 467.03	14	1 416 463 851.63	17
WV	52 573 250.16	27	1 279 700 567.46	16	1 332 273 817.62	18
KY	95 853 451.62	18	901 321 017.96	19	997 174 469.58	19
MI	71 926 313.66	22	881 918 999.84	20	953 845 313.50	20
WA	5 810 833.44	39	831 868 039.88	21	837 678 873.32	21
KS	242 094 678.50	11	579 467 561.02	24	821 562 239.52	22
NV	6 691 883.36	37	746 793 333.37	22	753 485 216.73	23
ME	510 000.06	46	701 426 205.16	23	701 936 205.22	24
MO	137 230 899.81	16	526 240 196.80	26	663 471 096.61	25
CO	10 705 500.03	35	547 930 000.02	25	558 635 500.05	26
AR	80 880 027.74	21	433 009 190.86	27	513 889 218.60	27
NC	141 781 900.43	15	369 230 638.53	31	511 012 538.96	28
AZ	185 044 355.91	14	322 225 200.52	32	507 269 556.43	29
GA	54 202 911.65	25	416 366 840.75	29	470 569 752.40	30
UT	51 514 299.95	28	403 347 000.27	30	454 861 300.22	31
OR	17 985 600.12	32	421 172 472.98	28	439 158 073.10	32
NE	219 264 314.90	12	199 736 914.82	37	419 001 229.72	33
OK	134 920 710.29	17	263 840 686.68	34	398 761 396.97	34
VT	10 950 999.96	34	277 911 120.48	33	288 862 120.44	35
TN	44 005 049.82	29	229 669 386.81	35	273 674 436.63	36
SD	85 762 177.01	19	142 213 569.61	42	227 975 746.62	37
MD	2 354 360.87	42	217 430 419.37	36	219 784 780.24	38
HI	3 885 200.00	41	186 869 550.03	38	190 754 750.03	39
MA	6 150 000.09	38	168 897 002.65	39	175 047 002.74	40
AK	17 131.26	50	161 935 531.00	40	161 952 662.26	41
CT	5 700 000.00	40	149 607 498.01	41	155 307 498.01	42
ID	1 503 000.01	43	126 247 500.04	43	127 750 500.05	43
SC	18 907 234.67	31	107 158 336.93	44	126 065 571.60	44

Table 2.3 (*cont.*)

	Crop damage		Property damage		Crop and property damage	
State	Amount	Rank	Amount	Rank	Amount	Rank
NM	13 767 500.17	33	96 234 550.12	45	110 002 050.29	45
NH	250 000.05	48	71 447 200.04	46	71 697 200.09	46
WY	747 505.02	45	64 969 400.05	47	65 716 905.07	47
DE	500 000.01	47	46 474 001.03	48	46 974 001.04	48
MT	9 356 044.99	36	32 934 545.11	49	42 290 590.10	49
DC	71 739.13	49	29 927 579.71	50	29 999 318.84	50
RI	0.00	51	7 796 000.00	51	7 796 000.00	51
Total	11 383 059 595.35		52 353 885 556.20		63 736 945 151.55	

Adapted from Hazards Research Lab: SHELDUS:(http://webra.cas.sc.edu/hvri/products/sheldus.aspx; accessed February 1, 2010).

flood losses over time are most likely a combination of risk exposure, social vulnerability, and urban and suburban growth in low-lying areas.

Trends for insured property damage at national and state levels

Insured losses catalogued through the NFIP offer another window through which we can view the magnitude of flood damage across the U.S. The spatial and statistical patterns of these data are different than for SHELDUS, but perhaps even more important when assessing the economic burden of floods on governments and local taxpayers.

Figure 2.8 depicts the overall trend for insured flood losses (focusing on property damage) in real dollars between 1978 and 2007. FEMA insurance claim data show that the number of policies in force increased 3.8 times during this time period, and average annual losses expanded from $147 million in 1978 to $17 billion in 2005. Insured flood losses are usually lower than losses reported in other data sources because they account only for a portion of the overall economic impact. However, the NFIP reported over $17 billion of payments to policy holders in 2005, a huge spike compared with previous years. This figure is much higher than other flood-based data sources for that year because the NFIP includes coastal flooding caused by storm surges. For example, when Hurricane Katrina hit the U.S. coast in 2005, the NFIP included the property damage from coastal flooding resulting from the storm-surge in

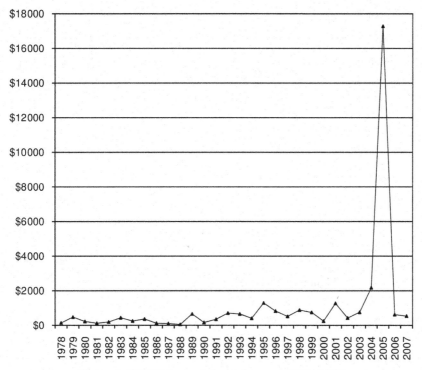

Figure 2.8 NFIP insured flood damage, 1978–2007.

the Gulf of Mexico. Other agencies establish a different damage category for hurricane-induced inundation.

To investigate the status and trends for insured property loss in more detail, we focus on the time period from 1996 to 2007 because more communities had federal insurance policies, and record keeping was likely more accurate. During these 12 years, we found that NFIP paid 623 220 claims totaling over $26 billion. The NFIP average annual loss from floods during this time period was approximately $2 billion, with an average payment per claim of $42 257.

The NFIP paid over 211 000 claims totaling over $17 billion in 2005 because of major coastal flooding events (e.g. Hurricanes Dennis, Katrina, Rita, Wilma, and Tropical Storm Tammy) that battered coastal communities that year. In fact, property damage from floods during 2005 amounted to over 66% of the total insured losses from 1996 to 2007 (Figure 2.9). Hurricane Katrina, the most devastating disaster in U.S. history, generated 165 618 insurance claims with total payments of over $16 billion and an average claim payment of $95 813. It seems, up until 2007,

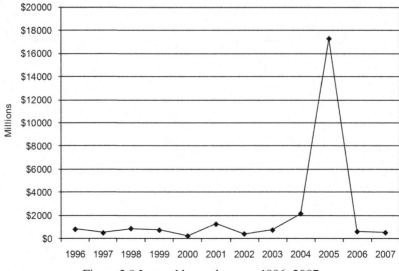

Figure 2.9 Insured losses by year, 1996–2007.

Table 2.4 *Top 10 and bottom 10 U.S. states in accumulated insured flood loss,*
1996–2007

Rank	State	Number of claims	Building	Contents	Increased cost of compliance	Total insured losses
1	Louisiana	178 330	10 908 375 062	2 866 401 834	96 372 048	13 871 148 944
2	Mississippi	26 169	1 988 274 911	584 913 214	19 303 790	2 592 491 916
3	Florida	90 292	1 940 391 403	492 264 970	17 649 393	2 450 305 765
4	Texas	63 665	1 431 653 839	451 898 482	8 587 609	1 892 139 931
5	Alabama	18 906	651 389 126	133 031 707	8 653 179	793 074 012
6	North Carolina	35 056	556 420 522	96 200 694	10 066 026	662 687 242
7	Pennsylvania	28 778	499 760 362	140 541 990	5 792 382	646 094 734
8	New Jersey	24 085	372 688 284	107 962 810	1 330 972	481 982 065
9	Virginia	18 204	293 045 906	51 527 900	14 535 561	359 109 367
10	New York	16 114	241 483 879	53 200 937	924 696	295 609 512
45	Colorado	487	4 445 135	568 528	55 802	5 069 465
46	Nebraska	415	3 851 948	761 607	0	4 613 555
47	Idaho	286	3 514 591	640 324	0	4 154 916
48	Vermont	264	3 033 284	518 491	0	3 551 775
49	Montana	433	2 915 458	285 624	0	3 201 082
50	Alaska	107	1 572 111	136 827	20 000	1 728 938
51	Guam	90	1 148 600	342 856	0	1 491 456
52	District of Columbia	42	1 252 816	92 371	0	1 345 187
53	Utah	62	441 732	48 088	0	489 820
54	Wyoming	39	322 212	1486	0	323 699

Adapted from Hazards Research Lab: SHELDUS (http://webra.cas.sc.edu/hvri/products/
sheldus.aspx; accessed February 1, 2010).

insured flood damage to residential property in the U.S. was dominated by a single catastrophic event: Hurricane Katrina.

Examining the record of NFIP losses by state illustrates the spatial pattern of vulnerability to floods across the U.S. Table 2.4 reports insured property damage estimates for the top and bottom 10 states from 1996 to 2007. NFIP-based insurance provides three types of coverage: building, contents, and increased cost of compliance (ICC). ICC coverage applies to a building which has been declared substantially damaged or repetitively damaged such that there are increased costs to comply with state or community floodplain management laws or ordinances after a flood. Generally, ICC helps pay for the cost of building elevation, relocation, demolition, or flood-proofing. This coverage can be provided in addition to building or contents coverage for actual physical damage, but it cannot exceed $20000 (FEMA, 2002). The category of total insured loss represents the sum of all three types of payment.

Due to Hurricane Katrina in 2005, Louisiana reported the largest amount of property damage (over $13 billion), followed by Mississippi (over $2 billion), Florida (over $2 billion), Texas (over $1.8 billion), and Alabama (over $793 million). In contrast, the District of Columbia, Utah, and Wyoming incurred the least flood losses. Insured losses were less than $1 million in Wyoming and Utah over the 12-year period from 1996 to 2007.

Figure 2.10 shows most clearly the areas of high vulnerability to flood damages across the U.S. The Gulf of Mexico coastline suffered the most intense losses, led by Louisiana, which took the brunt of Hurricane Katrina's landfall in 2005. Total insured losses for Louisiana were over five times that of the second most damaged state, Mississippi, and approximately 43000 times that of Wyoming. Several states along the mid-Atlantic coast, such as North Carolina and Pennsylvania, also reported large amounts of losses during the study period. By contrast, the central and northwest parts of the country sustained significantly less damage from floods. Flood damages are driven by multiple factors (as investigated in detail in Chapter 7), including low-lying, storm-prone areas and rapid population growth. Communities in these highly vulnerable areas must cope with a constant stream of repetitive floods, punctuated by major storm events bringing large amounts of rainfall and wave action to affected areas. The record of significant flood events from 1996–2007, as listed in Table 2.5, indicates the economic havoc caused by both tropical and non-tropical events in coastal communities.

Overall, the record of property damage caused by floods varies by year and across space. Rare, catastrophic events, such as Hurricane Katrina (the costliest storm in recorded history) often tip the scales of economic impact, depending on the time period being assessed. Small-scale events may go unnoticed among the

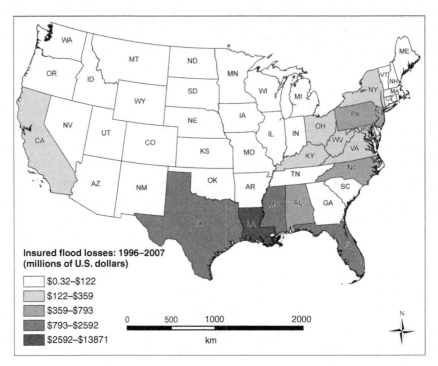

Figure 2.10 Insured flood losses by state, 1996–2007.

Table 2.5 *Significant flood events in the U.S., 1996–2007*

Year	Name of event	Date	No. of paid policy	Amount of losses	Average paid loss
1996	Northwest Flood	Feb–96	2329	$61 903 974	$26 580
	Bertha	Jul–96	1166	$10 388 364	$8 909
	Fran	Sep–96	10 315	$217 844 647	$21 119
	Hortense	Sep–96	1381	$20 215 202	$14 638
	Josephine	Oct–96	6512	$102 604 272	$15 756
	Northeast Flood	Oct–96	3480	$40 837 392	$11 735
	California Flood	Dec–96	1858	$39 697 267	$21 366
1997	South Central Flood	Feb–97	4529	$100 436 961	$22 176
	Upper Midwest Flood	Apr–97	7398	$160 102 096	$21 641
1998	Pineapple Express	Jan–98	4228	$57 677 068	$13 642
	Nor'Easter	Feb–98	3212	$28 011 723	$8721
	Hurricane Bonnie	Aug–98	2675	$23 073 621	$8626
	Texas Flood	Sep–98	4876	$78 402 450	$16 079

Table 2.5 (*cont.*)

Year	Name of event	Date	No. of paid policy	Amount of losses	Average paid loss
	Louisiana Flood	Sep–98	5174	$50 987 804	$9855
	Hurricane Georges (Keys)	Sep–98	3436	$43 134 378	$12 554
	Hurricane Georges-MS,PR,LA	Sep–98	848	$14 150 532	$16 687
	Hurricane Georges (Panhandle)	Sep–98	1680	$23 250 392	$13 840
	Texas Flood	Oct–98	3190	$143 580 854	$45 010
1999	Hurricane Floyd	Sep–99	20 439	$462 270 253	$22 617
	Hurricane Irene	Oct–99	13 682	$117 922 109	$8619
2000	Florida Flood	Oct–00	9276	$158 283 182	$17 064
2001	Tropical Storm Allison	Jun–01	30 662	$1 103 765 221	$35 998
	Tropical Storm Gabrielle	Sep–01	2418	$34 836 088	$14 407
2002	Texas Flood	Jul–02	1896	$70 634 069	$37 254
	Tropical Storm Isadore	Sep–02	8442	$113 691 962	$13 467
	Hurricane Lili	Oct–02	2563	$36 900 365	$14 397
	Texas Flood	Oct–02	3250	$88 984 769	$27 380
2003	Hurricane Isabel	Sep–03	19 852	$491 649 350	$24 766
2004	Hurricane Charley	Aug–04	2608	$50 607 681	$19 405
	Hurricane Frances	Sep–04	4952	$151 454 257	$30 584
	Hurricane Ivan	Sep–04	27 574	$1 571 160 291	$56 980
	Hurricane Jeanne	Sep–04	5373	$127 303 899	$23 693
2005	Hurricane Dennis	Jul–05	3795	$118 898 101	$31 330
	Hurricane Katrina	Aug–05	166 464	$16 016 992 444	$96 219
	Hurricane Rita	Sep–05	9463	$462 565 949	$48 882
	Tropical Storm Tammy	Oct–05	4 116	$44 728 148	$10 867
	Hurricane Wilma	Oct–05	9 597	$362 866 548	$37 810
2006	PA, NJ, NY Floods	Jun–06	6 403	$226 150 757	$35 319
	Hurricane Paul	Oct–06	1 507	$37 233 617	$24 707
2007	Nor'Easter	Apr–07	8 623	$224 651 554	$26 053

Source: FEMA (www.fema.gov/business/nfip/statistics/sign1000.shtm; accessed February 1, 2010).

barrage of statistics, but result in significant cumulative impacts over many years. Out of the multitude of maps, figures, claims, and pay-out estimates, the Gulf of Mexico coast and its surrounding jurisdictions emerges as a hotspot of negative impacts from floods. In particular, Texas and Florida are perennial stand-outs for human casualties, property damage, and the sheer number of flooding events over time. In the next chapter we look more closely at flood impacts within these two states, which serve as the focus areas for the analyses in the remainder of the book.

3

Impacts of flooding in coastal Texas and Florida

A glance at national and statewide flooding trends as presented in the previous chapter reveals just how vulnerable Texas and Florida are, compared with other states in the U.S. Both states consistently rank in the upper echelon when it comes to financial and human losses associated with floods. In fact, Texas experiences so many human casualties from flooding, the figure presents itself as a statistical outlier in our datasets. Moreover, Texas ranked fourth in the U.S. in property damages (both overall and insured) and sixth in crop damages. Florida fared no better in our analysis. This state was thirteenth in overall property damages (third for insured losses at almost $2.5 billion) and fourth in crop damages from floods.

These figures should come as no surprise. Both Texas and Florida contain very large populations (second and fourth in the U.S., respectively) living in vulnerable coastal areas, with the population expected to increase in the near future. Rapid population growth and sprawling development in low-lying, flood-prone areas that receive large amounts of yearly rainfall is a basic recipe for creating flood disasters. These states have set themselves up in a development–disaster cycle. However, Texas and Florida are large and varied, and floods tend to be localized problems. To really understand the nature of exposure to risk and flooding in Texas and Florida, we need to magnify our spatial resolution and better understand the variations within the states themselves.

In this chapter, we "drill down" below state boundaries to investigate the temporal–spatial patterns of flooding at the local level. In doing so, we address two research questions: (1) just how vulnerable are Texas and Florida, and (2) precisely where are the hotspots of vulnerability and flood loss within our study states? This analysis brings us one step closer to understanding where and why local communities experience increasing property damage and human casualties from floods.

Physical vulnerability to floods in coastal Texas and Florida

A major factor contributing to the high degree of flood vulnerability in Texas and Florida is the sheer number of people living in coastal zones. Coastal natural amenities, combined with promising economic opportunities, make these areas magnets for population growth, thereby exposing more people, structures, and property to the adverse impacts of floods. Harris County and the greater Houston area contain the greatest number of people living along the Texas coast, with almost 2000 people per square mile calculated for the year 2000. The population in Harris County was expected to increase by more than 60% from 2003 to 2008, making it the most populous in the state (Crossett *et al.*, 2004). Overall, five of the top 20 most populated counties in Texas (Harris, Cameron, Brazoria, Galveston, and Jefferson counties) are located on the coast. In 2000, over 36% (7.5 million) of the Texas population resided in 54 coastal counties at significantly higher densities (146 persons per square mile) compared to the rest of the state (79 persons per square mile) and the U.S. (80 persons per square mile) (see Table 3.1).

Population growth along the Florida coast has been even more extensive over the past several decades, with significant increases in density in counties with coastlines. In the year 2000, over 12 million people were living in 35 counties fringing the Gulf of Mexico and Atlantic Ocean, which accounted for about 77% of the entire Florida population (Table 3.1). Population density in these areas exceeded 400 persons per square mile, with Pinellas (over 92% built-out), Broward, and Miami–Dade counties taking the lead.

Vulnerability to persistent floods is not simply a function of a growing number of people in coastal areas, but exactly where they choose to live, work, and play. The 100-year floodplain (where there is a 1% chance of flooding each year) has long been considered the most vulnerable part of a coastal landscape within which to place structures. In fact, over the course of a 30-year mortgage, there is a 26% chance that a home in the floodplain will experience flooding (National Research Council [NRC], 2000). FEMA delineates, documents, and publishes floodplain boundaries available to the general public, including prospective homebuyers, title companies, and real-estate agents. The decision to buy a home within the 100-year floodplain in low-lying coastal areas is often one of the differences between experiencing flood damage and remaining unaffected during storm events.

Our 54-county focal area in Texas comprises, on average, approximately 21% of the 100-year floodplain (see Appendix 3.1, Table A3.1.1 for a break-down of each county). Generally, as one moves closer to the coastline, the percentage of floodplain increases (see Figure 3.1). The problem is that counties adjacent to the Gulf of Mexico are targets for future population growth and associated structural development. Jefferson County contains the most floodplain area, with nearly 60% of its land base within this designation. Chambers and Brazoria counties are also highly

Table 3.1 *Population trends in coastal areas*

	1980		1990		2000		Land area (square miles)
	Population, n (%)	Population density[a]	Population, n (%)	Population density[a]	Population, n (%)	Population density[a]	
Texas coastal counties	5 203 957 (36.57)	101.32	6 070 491 (35.74)	118.19	7 513 502 (36.03)	146.28	51 362.77 (19.62)
Texas state	14 229 191 (100)	54.35	16 986 510 (100)	64.88	20 851 820 (100)	79.60	261 797.12 (100)
Florida coastal counties	7 692 909 (78.93)	256.68	10 096 282 (78.04)	336.87	12 285 697 (76.87)	409.95	29 971.19 (55.58)
Florida state	9 746 961 (100)	180.74	12 937 926 (100)	239.92	15 982 378 (100)	296.40	53 926.82 (100)
U.S.	226 542 199	64.04	248 709 873	70.31	281 421 906	79.56	3 537 438.44

a Population density per square mile of land area.

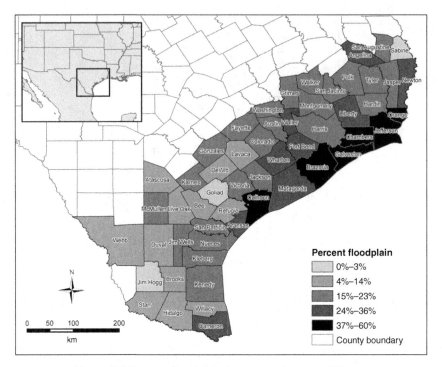

Figure 3.1 Percent floodplain by county in coastal Texas.

vulnerable places to build, with approximately half of their land areas designated as floodplain.

By comparison, Florida contains even more land area in the 100-year floodplain, with an average of 37% across all 67 counties (see Appendix 3.1, Table A3.1.2). In fact, over 22% of counties statewide have over half of their area designated as floodplain. As shown in Figure 3.2, Monroe and Miami–Dade counties, which are most directly associated with the Everglades ecosystem, are composed primarily of floodplains. Another hotspot of flood vulnerability is located in the Panhandle of the state: Franklin, Gulf, and Lafayette counties. Interestingly, this area is slated for large amounts of residential development over the next decade.

Flood impacts in coastal Texas

Casualties

As already noted in Chapter 2, Texas experiences by far more deaths and injuries from flooding events than any other state in the U.S. This chapter investigates precisely where and when within our 54-county coastal study area these incidents occur. As shown in Figure 3.3, over the last 50 years, Harris County (home to the city of Houston) has had the largest number of fatalities from floods. In contrast,

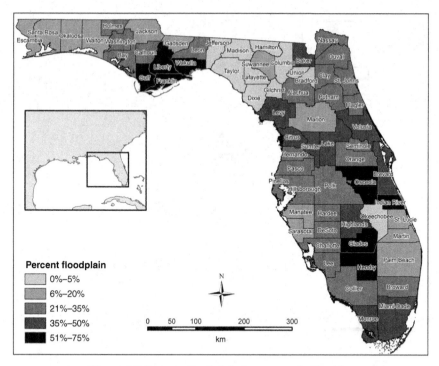

Figure 3.2 Percent floodplain by county in Florida.

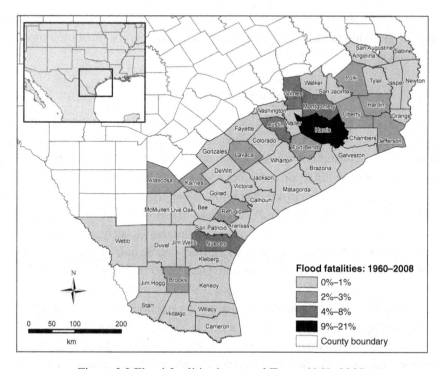

Figure 3.3 Flood fatalities in coastal Texas, 1960–2008.

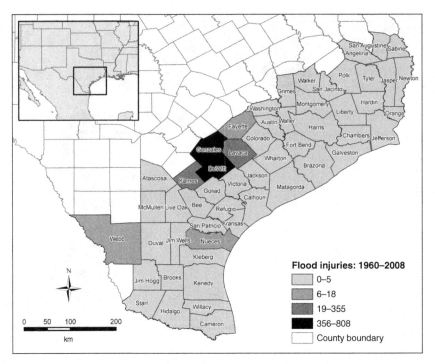

Figure 3.4 Flood injuries in coastal Texas, 1960–2008.

the roughly 2200 injuries during the same time period concentrate further inland among DeWitt, Gonzales, and Lavaca counties, which can be attributed mostly to the Guadalupe River flood in 1998 (Figure 3.4).

Crop and property damage

Texas ranks sixth in the nation for crop damage from floods between 1960 and 2008, with an estimated agricultural impact from flooding of $15 million per year. However, our 54-coastal county focal area received approximately $39 million of damage from 1960 to 2008, or only about 5% of the statewide total. As shown in Figure 3.5, most damage occurred in counties along the Mexican border and southern coastal bend, where agricultural operations are most prominent.

In comparison, property damage caused by floods is largely a coastal phenomenon in Texas. Our 54 coastal counties incurred almost half of the $2.6 billion of recorded property damage across the state from 1960 to 2008 (see Appendix 3.1, Table A3.1.3 for a breakdown of top counties). This figure translates into more than $23 million dollars each year in lost homes, schools, roads, and other structures. According to the data, Harris, Cameron, and Brooks counties received the most overall property damage (Figure 3.6) with over $100 million each. The drivers of flood damage are clearly different in each of these counties considering

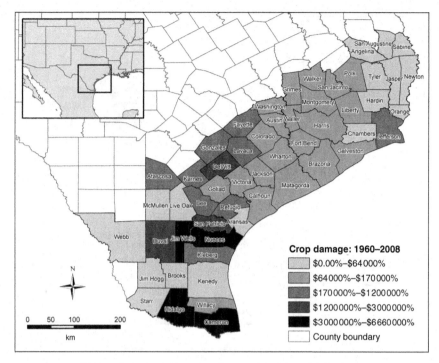

Figure 3.5 Crop damage of coastal Texas, 1960–2008.

that Harris has approximately four million residents while Brooks has only about 8000.

Insured property damage from floods

Coastal Texas played an even more prominent role in terms of insured property losses from floods. From 1996 to 2008, the NFIP paid over $1.5 billion to 362 participating communities to assist them in recovering from flooding events. This payout comes to almost 83% of the total losses covered statewide (Table 3.2). In 2001, when Tropical Storm Allison hit the greater Houston area, the effects were particularly severe with nearly $1 billion of insured losses representing over 97% of insured damage for all of Texas.

As shown in Table 3.3 and Figure 3.7, flood damage over the 12-year period was concentrated within Harris, Galveston, Jefferson, and Brazoria counties, where there was either a large number of people or a large percentage of floodplain. For example, Harris County experienced over $1 billion of property damage, which is over 500 times the amount in Hidalgo, the county ranked twentieth (Table 3.3). Galveston County, where almost half of the area is within the 100-year floodplain, is the second highest in property damage with almost $90 million. Even when controlling for population, the degree of insured flood damages was consistently

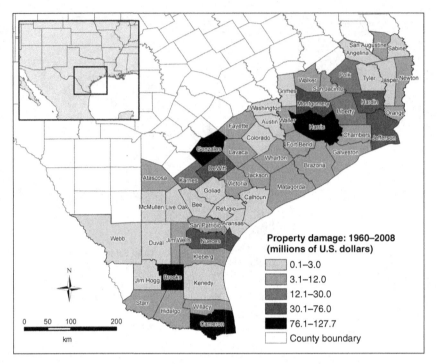

Figure 3.6 Property damage of coastal Texas, 1960–2008.

located in the Houston–Galveston region. Based on 2000 U.S. Census data, per capita damage for Harris and Galveston counties was approximately $341 and $358, respectively. In comparison, per capita insured damage over the same time period was only $4 for Hidalgo County and $156 for Aransas county.

Zip code level analysis of insured property damage

We refine the spatial resolution of our analysis one more time to examine insured flood damages in coastal Texas at the zip code level. Specifically, we analyzed 483 zip codes within our 54-county study area for the years 2006 and 2007. This rather unremarkable time period was chosen for analysis so as not to illustrate a biased portrait of flood loss in coastal Texas. Compared with other time periods, no extremely large or catastrophic flood events occurred. From 2006 to 2007, communities within the study area experienced almost $130 million in damages ($265 354 per zip code), which accounted for 70% of insured property damage for the entire state.

Thirty zip codes incurred flood damages over $1 million. As shown in Figure 3.8, these hotspots of vulnerability cluster primarily in the southeast portion of Houston along the Ship Channel and San Jacinto Bay. La Porte, Deer Park, and Pasadena are local communities near the industrial complex of the Houston area that received among the most damage. The zip code receiving the highest amount of flood damage was

Table 3.2 *Insured flood damage for Texas coastal counties and TX*

Year	Insured flood damage	
	54 coastal counties, $ (%)	Texas, £ (%)
1996	9 000 919.42 (88.75)	10 141 679.24 (100)
1997	23 645 898.74 (64.19)	36 837 177.53 (100)
1998	153 481 489.71 (57.85)	265 288 859.99 (100)
1999	7 367 063.06 (84.40)	8 728 356.74 (100)
2000	9 098 073.44 (61.87)	14 705 380.38 (100)
2001	973 619 092.29 (97.25)	1 001 132 975.12 (100)
2002	128 788 566.76 (64.85)	198 588 939.17 (100)
2003	52 097 617.17 (98.03)	53 142 829.26 (100)
2004	21 404 231.88 (35.21)	60 791 060.27 (100)
2005	55 872 700.26 (97.52)	57 293 601.71 (100)
2006	89 665 389.47 (90.49)	99 088 638.00 (100)
2007	39 633 604.29 (45.87)	86 400 433.16 (100)
Total	1 563 674 646.49 (82.64)	1 892 139 930.57 (100)

in Houston proper with over $8.43 million during the two-year period. This zip code is characterized by predominantly lower-income Hispanic populations configured in what is an average density for that region. While this socioeconomic composition suggests issues of social vulnerability, an examination of the top 10 flood-damaged zip codes shows no consistent pattern (see Appendix 3.1, Table A3.1.4). In fact, these local hotspots of vulnerability are comprised of mostly white residents with above-average incomes living in low-density spatial configurations.

Flood impacts in Florida

Casualties

Despite having more people and structures situated in floodplains than coastal Texas, Florida has historically been a much safer place to live when it comes to flood dangers. From 1960 to 2008, 23 people died and only eight were injured from

Table 3.3 *Top 20 Texas counties for insured flood loss, 1996–2007*

Rank	County	Total insured property damage, $
1	Harris	1 162 105 186.43
2	Galveston	89 566 835.94
3	Jefferson	73 055 580.39
4	Brazoria	72 070 514.42
5	Montgomery	50 997 112.47
6	Orange	18 148 750.77
7	Hardin	12 396 616.69
8	Kleberg	9 740 907.79
9	Liberty	8 608 526.61
10	Wharton	8 092 974.96
11	Matagorda	7 172 445.91
12	Cameron	6 425 194.70
13	Gonzales	5 819 973.43
14	Victoria	5 793 990.54
15	Fort Bend	3 919 011.84
16	DeWitt	3 820 836.77
17	Chambers	3 728 331.22
18	Nueces	3 590 616.67
19	Aransas	3 517 671.81
20	Hidalgo	2 264 454.20

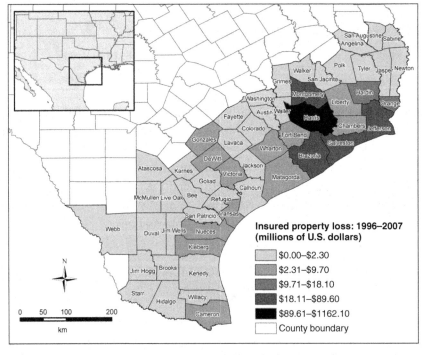

Figure 3.7 Insured property damage in coastal Texas, 1996–2007.

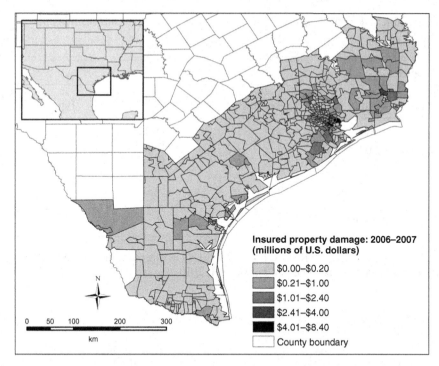

Figure 3.8 Insured property damage in Texas coastal zip codes, 2006–2007.

flooding events across the entire state (67 counties). As shown in Table 3.4, Duval County, which encompasses the city of Jacksonville, was the most dangerous local-ity with four fatalities (compared with 21 deaths in Harris County, TX); Pinellas and Hillsborough counties had the most numbers of flood-related injuries with only two each over a 48-year period.

Crop damage and property damage

While Florida was thirty-eighth among all states in the country for human cas-ualties, it ranked fourth in crop damages from floods. From 1960 to 2008, Florida incurred approximately $934 million in agricultural losses, most of which occurred in the years 1999 and 2000. Crop damage was concentrated in the southeastern part of the state in and around Miami–Dade, Broward, and Palm Beach Counties, areas with high concentrations of citrus and tropical fruit production (Figure 3.9).

Florida also endured large amounts of overall property damage from floods. From 1960 to 2008, the state racked up over $1.4 billion in damages, most of which occurred starting in the late 1990s. Once again, the southeastern-most counties led the way in financial losses, with Miami–Dade and Broward counties accounting for $599 million of total damages. The Panhandle area containing Leon and Gulf

Table 3.4 *Top counties for flood-related casualties and injuries in*
Florida, 1960–2008

Fatalities			Injuries		
Rank	County	Fatalities	Rank	County	Injuries
1	Duval	4	1	Hillsborough	2
2	Leon	3	2	Pinellas	2
3	Hillsborough	3	3	Broward	1
4	Manatee	3			
5	Hernando	2			
6	Pasco	2			
7	Pinellas	1			

Adapted from Hazards Research Lab: SHELDUS (http://webra.cas.sc.edu/hvri/products/sheldus.aspx; accessed February 1, 2010).

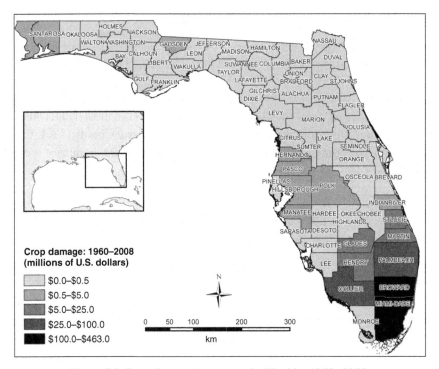

Figure 3.9 Crop damage by county in Florida, 1960–2008.

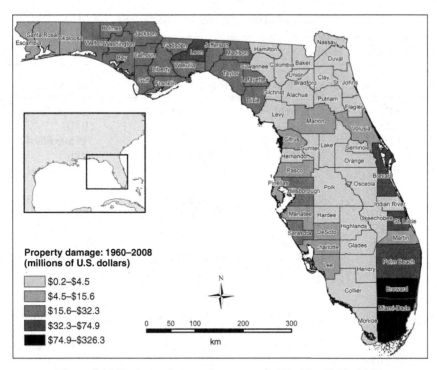

Figure 3.10 Property damage by county in Florida, 1960–2008.

counties is another hotspot for property damages caused by floods, even though most of the population is concentrated in the south (Figure 3.10 and Appendix 3.1, Table A3.1.5). Since the 1960s, property damage from floods has greatly increased, along with population growth and development on the coast. For example, the average recorded damage in the 1960s was approximately $8.54 million per year (inflation adjusted to 2000 real dollars); in the 2000s this figure jumped to over $72 million per year.

Insured flood damage

Unlike Texas, approximately 80% of Florida residents live or work on or near coastlines. As a consequence, 95% of all Florida communities participate in the NFIP, making insured losses a major factor associated with flood impacts. As of 2007, over 2.1 million federal insurance policies were issued in Florida – almost 41% of the total policies in the U.S. (Florida Division of Emergency Management, 2008).

From 1996 to 2007 alone, Florida residents filed more than 14% of all the claims in the U.S. and suffered almost 10% of the total damage from floods (Table 3.5). Actual insured flood losses during this time period total over $2.45 billion (about $200 million per year), which suggest overall damage estimates in the SHELDUS

Table 3.5 *Trends of insured flood loss in Florida and U.S.*

	Number of claims		Insured loss	
Year	Florida, n (%)	U.S., n (%)	Florida, $ (%)	U.S., $ (%)
1996	6813 (12.93)	52 679 (100)	105 329 118.67 (12.72)	827 790 157.25 (100)
1997	1478 (4.87)	30 338 (100)	12 101 397.00 (2.33)	519 505 659.47 (100)
1998	7918 (13.81)	57 350 (100)	101 693 042.33 (11.48)	886 112 489.15 (100)
1999	15 637 (33.10)	47 245 (100)	137 909 784.11 (18.27)	754 763 257.36 (100)
2000	10 157 (62.08)	16 361 (100)	166 664 170.24 (66.21)	251 711 107.99 (100)
2001	3330 (7.64)	43 560 (100)	45 368 736.86 (3.56)	1 273 664 923.02 (100)
2002	759 (3.00)	25 287 (100)	8 376 332.33 (1.94)	430 750 921.70 (100)
2003	1125 (3.06)	36 716 (100)	14 005 229.38 (1.84)	760 686 136.99 (100)
2004	22 075 (39.65)	55 668 (100)	1 220 916 286.20 (56.10)	2 176 325 247.19 (100)
2005	20 076 (9.51)	211 019 (100)	621 312 044.71 (3.59)	17 283 465 887.48 (100)
2006	524 (2.14)	24 458 (100)	10 295 708.93 (1.64)	627 074 582.73 (100)
2007	400 (1.79)	22 305 (100)	6 333 914.68 (1.16)	543 789 648.63 (100)
Total	**90 292** (14.49)	**623 220** (100)	**2 450 305 765.44** (9.30)	**26 335 640 018.96** (100)

database are grossly underestimated. The years 2000 and 2004 were particularly bad for insured flood losses. In 2000, for example, Florida incurred $166 million in losses, which was almost 67% of flood damages for the entire U.S. during that year. In 2004, the insured flood loss in Florida exceeded $1 billion due to several hurricanes – Charley, Frances, Ivan, and Jeanne – amounting to more than 56% of all losses nationwide.

Of the 368 Florida communities participating in the NFIP, residents in Santa Rosa County in the far northwest of the state claimed the largest amount of insured damage from 1996 to 2007, with over $358 756 000 (Table 3.6 and Figure 3.11). Neighboring Escambia County experienced the second highest amount of flood

Table 3.6 *Top 20 Florida counties for insured flood loss, 1996–2007*

Rank	Community	Total insured damage (1996–2007), $
1	Santa Rosa County	358 756 314.25
2	Escambia County	308 462 517.61
3	Miami–Dade County	242 016 493.76
4	City of Key West	169 768 976.56
5	Monroe County	168 555 134.48
6	Pensacola Beach-Santa Rosa Island Authority	137 639 644.95
7	Okaloosa County	86 111 111.80
8	City of Miami	54 391 475.53
9	Lee County	50 569 846.33
10	City of Destin	33 862 328.56
11	City of Marathon	33 799 119.08
12	St. Lucie County	31 516 653.01
13	Walton County	30 031 646.60
14	City of St. Petersburg	29 827 920.66
15	City of Pensacola	28 819 799.57
16	City of Vero Beach	27 043 237.71
17	City of Fort Pierce	25 406 630.96
18	City of Gulf Breeze	24 530 956.25
19	Indian River County	23 691 461.87
20	Martin County	22 679 676.34
Mean		94 374 047.29

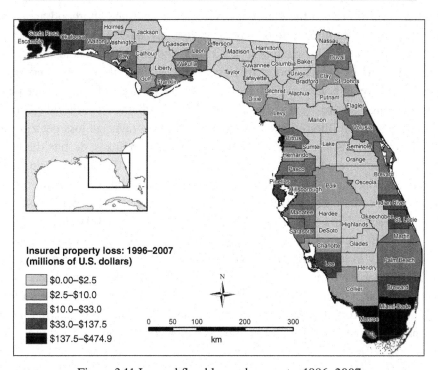

Figure 3.11 Insured flood losses by county, 1996–2007.

damage in the state with over $300 million. Miami–Dade and Key West in south-east Florida also ranked in the top five places for insured damages.

In all, the top 20 communities represented 77% of the total damages statewide (Table 3.6). The per capita average flood loss in these 20 communities was approximately $3141, which is 20 times the per capita average for the state. The majority of the communities are located in flood-prone areas either directly on the coast or on islands. For example, the Key West, Key Colony Beach, Layton, Marathon, and Monroe County cluster on the southern Florida peninsula in areas vulnerable to storms and flooding all claimed more than $1400 per person in flood damages from 1996 to 2007. Another noteworthy cluster of coastal communities vulnerable to per capita losses is situated adjacent to Pensacola and Choctawhatchee Bays in the northwest of the Florida peninsula. This area includes Santa Rosa County and the cities of Gulf Breeze, Destin, and Shalimar. In contrast, inland communities incurred significantly lower amounts of damage at all spatial scales. While flood damages tend to increase with population, it is interesting to note that this is not always the case. For example, only two counties (Monroe County and Santa Rosa County) are ranked among the top 20 communities for both overall and per capita insured damages.

Zip code-level analysis of insured property damage (2006–2007)

A spatial "drill-down" to the zip code level shows local neighborhood areas that are particularly susceptible to flood damage. We analyzed 930 zip codes across the 67-county statewide study area for the years 2006 and 2007 to find hotspots of flood vulnerability. As explained above, these years represent fairly typical amounts of storm-related precipitation so as not to show a biased picture of flood damage punctuated by a single severe event. From 2006 to 2007, communities in Florida reported approximately $17 million in insured property damages (average loss per zip code was $17 605). This is a fraction of the amount reported in Texas, despite the larger Florida study area and higher participation rates in the NFIP. Only two zip codes, one in Palm Beach and the other in Miami, submitted insurance claims for over $1 million (a zip code in Palm Beach had the highest amount of loss with $1.45 million). Thirty-seven zip codes reported flood damages of over $100 000.

As shown in Figure 3.12, several spatial hotspots of flood loss exist across the state, primarily in low-lying, urbanized coastal areas. Multiple adjacent zip codes with high amounts of insured flood damage are especially pronounced in the Miami Beach area in the southeast part of the state and Jacksonville to the north. A hotspot of flood property loss also emerges on the west coast of the state in Pinellas Park and St. Petersburg along Tampa Bay, as well as further up the coast in the less

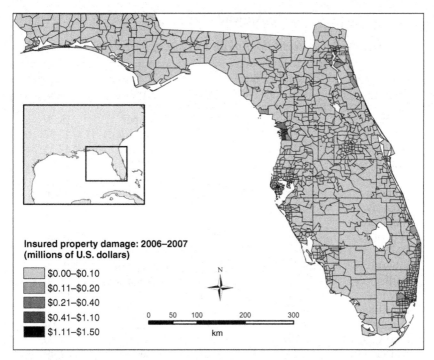

Figure 3.12 Insured property damage in Florida zip codes, 2006–2007.

populated communities of Homosassa and Crystal River. As is the case with Texas, the socioeconomic makeup of the most flood-damaged zip codes varies widely. These areas are represented by diversity of income levels, race, population densities, percentage of renters, etc. (see Appendix 3.1, Table A3.1.5).

Summary

In Chapter 2, we showed that Texas and Florida are among the states that are most susceptible to the adverse impacts of floods in the U.S. In this chapter, we demonstrate through a multiscale analysis that specific areas within each state are more vulnerable to flood damage than others. Given the variation of conditions within states, it is important to conduct assessments at the finest resolution possible, even below the county level where most data are collected. Results indicate hotspots of flood damage in mostly urbanized localities adjacent to or in very close proximity to the coast. However, no clear pattern emerges as to why specific areas receive significantly more damage than others, and how this damage can be mitigated in the future. These are precisely the issues we address in the following chapters.

The consequences of floods

Appendix 3.1

Table A3.1.1 *Percent floodplain by county in coastal Texas*

Order	County	Land area in flood zone, %	Order	County	Land area in flood zone, %
1	Jefferson	60.03	28	Jasper	18.29
2	Chambers	50.73	29	Jim Wells	17.57
3	Brazoria	48.29	30	Fayette	17.55
4	Galveston	46.47	31	Polk	17.40
5	Orange	46.39	32	Jackson	17.37
6	Calhoun	42.35	33	San Patricio	17.16
7	Aransas	36.29	34	Colorado	17.13
8	Matagorda	33.81	35	Walker	17.00
9	Wharton	32.51	36	Tyler	16.34
10	Cameron	31.98	37	Harris	16.17
11	Liberty	29.95	38	Grimes	16.02
12	Waller	25.19	39	Karnes	15.56
13	Newton	24.77	40	DeWitt	14.28
14	Angelina	24.57	41	Hidalgo	13.88
15	Fort Bend	24.57	42	Refugio	13.84
16	San Augustine	22.79	43	Atascosa	13.20
17	San Jacinto	22.72	44	Bee	12.99
18	Victoria	21.85	45	Webb	12.95
19	Kennedy	21.65	46	Lavaca	12.84
20	Montgomery	20.77	47	Starr	11.34
21	Kleberg	20.72	48	Duval	10.58
22	McMullen	20.44	49	Willacy	10.21
23	Austin	19.10	50	Brooks	9.57
24	Washington	18.70	51	Live Oak	8.86
25	Hardin	18.58	52	Jim Hogg	2.74
26	Nueces	18.46	53	Goliad	0.07
27	Gonzales	18.43	54	Sabine	0.04

Average: 21.35%

Source: Hazards and Vulnerability Research Institute, University of South Carolina, *Oxfam America* (http://adapt.oxfamamerica.org/; accessed March 20, 2010).

Table A3.1.2 *Percent floodplain by county in Florida*

Order	County	Land area in flood zone, %	Order	County	Land area in flood zone, %
1	Monroe	98.11	35	Orange	32.70
2	Franklin	87.82	36	Lee	31.86
3	Gulf	76.26	37	Charlotte	31.42
4	Hendry	74.61	38	Holmes	30.53
5	Lafayette	68.66	39	Calhoun	30.05
6	Indian River	67.40	40	Alachua	29.47
7	Taylor	63.78	41	Hamilton	29.38
8	Dixie	63.16	42	Bay	29.04
9	Glades	62.39	43	Hillsborough	28.48
10	Wakulla	62.06	44	Leon	27.37
11	Liberty	58.60	45	Union	26.18
12	Brevard	53.30	46	Washington	26.09
13	Miami–Dade	52.84	47	Duval	25.31
14	Osceola	51.50	48	Clay	24.04
15	Volusia	51.17	49	Gilchrist	23.98
16	Okeechobee	47.66	50	DeSoto	23.72
17	Levy	47.23	51	Hernando	23.55
18	Sumter	44.40	52	Hardee	22.51
19	Lake	44.35	53	Martin	21.07
20	Polk	43.18	54	Collier	19.97
21	Bradford	41.08	55	Jackson	19.69
22	Jefferson	39.95	56	Walton	19.41
23	Madison	39.67	57	Sarasota	19.34
24	Baker	39.18	58	Manatee	18.91
25	St. Johns	38.13	59	Marion	18.70
26	Putnam	37.56	60	Broward	16.77
27	Highlands	37.36	61	Santa Rosa	16.33
28	Citrus	36.27	62	Escambia	16.21
29	Columbia	36.07	63	Okaloosa	14.02
30	Flagler	35.74	64	Suwannee	12.31
31	Pasco	34.82	65	Gadsden	11.12
32	Pinellas	34.50	66	Palm Beach	10.78
33	Seminole	32.85	67	St. Lucie	7.04
34	Nassau	32.78		Average = 36.89%	

Source: Hazards and Vulnerability Research Institute, University of South Carolina, *Oxfam America* (http://adapt.oxfamamerica.org/; accessed March 20, 2010).

Table A3.1.3 *Top 10 Texas coastal counties for crop and property damage,*
1960–2008

Rank	County	Crop damage, $	Rank	County	Property damage, $
1	Jim Wells	6 668 342.00	1	Harris	127 680 526.00
2	Nueces	5 358 342.00	2	Cameron	110 867 552.00
3	Cameron	5 077 786.00	3	Brooks	100 114 564.00
4	Hidalgo	5 000 008.00	4	Gonzales	90 878 035.00
5	DeWitt	2 914 515.00	5	DeWitt	75 982 368.00
6	San Patricio	2 425 009.00	6	Montgomery	58 354 682.00
7	Duval	1 668 342.00	7	Nueces	57 592 641.00
8	Gonzales	1 201 182.00	8	Hardin	53 914 744.00
9	Lavaca	1 167 348.00	9	Jefferson	31 517 638.00
10	Jefferson	1 002 791.00	10	Liberty	27 153 525.00

Source: Hazards Research Lab: SHELDUS (http://webra.cas.sc.edu/hvri/products/sheldus.
aspx; accessed February 1, 2010).

Table A3.1.4 *Top 10 most flood-damaged zip codes in coastal Texas*

Zip code: 77034	Insured damage: $8 437 236.71	Community name: Houston

- Population (2000): 26 342
- Population density: 3159 people per square mile
- No. of housing units: 10 312
- Median household income: $37 521
- Median house value: $64 200
- Renters: 55%
- Race: Hispanic 49.2%
- Poverty rate: 13%

Zip code: 77571	Insured damage: $7 145 254.03	Community name: La Porte

- Population (2000): 33 736
- Population density: 938 people per square mile
- No. of housing units: 12 461
- Median household income: $56 552
- Median housing value: $83 200
- Renters: 22%
- Race: white alone 82%
- Poverty rate: 7.2%

Zip code: 77536	Insured damage: $6 206 131.64	Community name: Deer Park

- Population (2000): 28 635
- Population density: 3248 people per square mile
- No. of housing units: 9978
- Median household income: $ 61 082

Table A3.1.4 (*cont.*)

- Median housing value: $90 700
- Renters: 21%
- Race: white alone 89.7%
- Poverty rate: 5.6 %

Zip code: 77087	Insured damage: $6 018 724.70	Community name: Houston

- Population(2000): 36 194
- Population density: 6236 people per square mile
- No. of housing units: 10 895
- Median household income: $30 346
- Median housing value: $53 200
- Renters: 40%
- Race: Hispanic 72.4%
- Poverty rate: 24.8%

Zip code: 77505	Insured damage: $5 183 610.79	Community name: Pasadena

- Population (2000): 20 395
- Population density: 2260 people per square mile
- No. of housing units: 7295
- Median household income: $60 403
- Median housing value: $107 800
- Renters: 31%
- Race: white alone 82.5%
- Poverty rate: 7.4 %

Zip code: 77657	Insured damage: $3 984 219.57	Community name: Lumberton

- Population (2000): 15 031
- Population density: 236 people per square mile
- No. of housing units: 5930
- Median household income: $ 41 202
- Median housing value: $ 89 600
- Renters: 18%
- Race: white alone 97.6%
- Poverty rate: 6.4%

Zip code: 77015	Insured damage: $3 342 814.09	Community name: Houston

- Population (2000): 50 569
- Population density: 2633 people per square mile
- No. of housing units: 17 513
- Median household income: $40 032
- Median housing value: $72 400
- Renters: 41%
- Race: Hispanic 44.3%
- Poverty rate: 15.8%

Table A3.1.4 (*cont.*)

Zip code: 77075   Insured damage: $3 293 562.42   Community name: Houston

---

- Population (2000): 22 423
- Population density: 2241 people per square mile
- No. of housing units: 8111
- Median household income: $38 188
- Median housing value: $69 700
- Renters: 45%
- Race: Hispanic 50.5%
- Poverty rate: 12.8%

Zip code: 77061   Insured damage: $3 082 232.83   Community name: Houston

---

- Population: 25 500
- Population density: 3573 people per square mile
- Housing units: 9465
- Median household income: $30 440
- Median housing value: $78 400
- Renters: 61%
- Race: Hispanic 47.9%
- Poverty rate: 18.3%

Zip code: 77033   Insured damage: $3 001 194.47   Community name: Houston

---

- Population: 27 676
- Population density: 5211 people per square mile
- No. of housing units: 9495
- Median household income: $26 544
- Median housing value: $36 300
- Renters: 31%
- Race: black or African American 84.3%
- Poverty rate: 31.3%

Source: U.S. Census Bureau (www.census.gov/; accessed November 17, 2010).

Table A3.1.5 *Top 10 Florida counties for crop and property damage, 1960–2008*

Rank	County	Crop damage, $	Rank	County	Property damage, $
1	Miami–Dade	463 000 138.89	1	Miami–Dade	326 254 515.79
2	Broward	275 000 138.89	2	Broward	273 450 441.72
3	Palm Beach	80 000 638.89	3	Brevard	74 894 441.05
4	Collier	37 000 138.89	4	St. Lucie	71 309 441.72
5	St. Lucie	20 000 138.89	5	Palm Beach	64 089 941.72
6	Hendry	15 000 000.00	6	Leon	47 087 829.78
7	Glades	15 000 000.00	7	Bay	32 330 292.75
8	Martin	9 000 138.89	8	Gulf	28 977 792.75

Table A3.1.5 *(cont.)*

Rank	County	Crop damage, $	Rank	County	Property damage, $
9	Indian River	5 000 138.89	9	Franklin	28 744 459.41
10	Manatee	1 732 281.75	10	Holmes	28 242 862.19

From Hazards Research Lab: SHELDUS (http://webra.cas.sc.edu/hvri/products/sheldus. aspx; accessed February 1, 2010).

Table A3.1.6 *Top ten most flood-damaged zip codes in coastal Florida*

Zip code: 33480   Insured damage: $1 453 298.03   Community name: Palm Beach

- Population (2000): 11 200
- Population density: 2608 people per square mile
- No. of housing units: 10 864
- Median household income: $ 84 191
- Median housing value: $1 000 000+
- Renters: 16%
- Race: white alone 96.2%
- Poverty rate: 6.0%

Zip code: 33166   Insured damage: $1 116 134.61   Community name: Miami

- Population (2000): 22 563
- Population density: 2148 people per square mile
- No. of housing units: 8561
- Median household income: $ 43 684
- Median housing value: $ 149 500
- Renters: 49%
- Race: Hispanic 62.1%
- Poverty rate: 11%

Zip code: 34448   Insured damage: $892 891.25   Community name: Homosassa

- Population (2000): 10 006
- Population density: 165 people per square mile
- No. of housing units: 5680
- Median household income: $ 28 443
- Median housing value: $ 87 800
- Renters: 16%
- Race: white alone 96.6%
- Poverty rate: 12.7%

Zip code: 33781   Insured damage: $802 602.52   Community name: Pinellas Park

- Population (2000): 25 287
- Population density: 3854 people per square mile
- No. of housing units: 11 417
- Median household income: $ 33 732
- Median housing value: $ 68 700

Table A3.1.6 (*cont.*)

- Renters: 34%
- Race: white alone 88.9%
- Poverty rate: 11.7%

Zip code: 33139   Insured damage: $782 157.25   Community name: Miami Beach

- Population (2000): 38 441
- Population density: 15 726 people per square mile
- No. of housing units: 29 094
- Median household income: $ 26 082
- Median housing value: $ 470 700
- Renters: 70%
- Race: Hispanic 50.7%
- Poverty rate: 22.7%

Zip code: 34429   Insured damage: $741 585.14   Community name: Crystal River

- Population (2000): 8575
- Population density: 196 people per square mile
- No. of housing units: 4885
- Median household income: $33 861
- Median housing value: $91 700
- Renters: 20%
- Race: white 92.8%
- Poverty Rate: 11.8%

Zip code: 33782   Insured damage: $583 413.07   Community name: Pinellas Park

- Population (2000): 19 527
- Population density: 3956 people per square mile
- No. of housing units: 9064
- Median household income: $ 37 049
- Median housing value: $ 78 500
- Renters: 15%
- Race: white 90.5%
- Poverty rate: 7.7%

Zip code: 33709   Insured damage: $391 082.66   Community name: St.Petersburg

- Population (2000): 26 039
- Population density: 4833 people per square mile
- No. of housing units: 14 067
- Median household income: $ 29 098
- Median housing value: $ 73 500
- Renters: 23%
- Race: white 90.3%
- Poverty rate: 11.3%

Table A3.1.6 (*cont.*)

Zip code: 34428   Insured damage: $329 803.07   Community name: Crystal River

- Population (2000): 9294
- Population density: 123 people per square mile
- Housing units: 4621
- Median household income: $30 069
- Median housing value: $88 600
- Renters: 22%
- Race: white 92.2%
- Poverty rate: 13.4%

Zip code: 33130   Insured damage: $312 519.34   Community name: Miami

- Population (2000): 20 541
- Population density: 22 085 people per square mile
- No. of housing units: 9107
- Median household income: $ 13 684
- Median housing value: $ 83 200
- Renters: 89%
- Race: Hispanic 90.5%
- Poverty rate: 38.1%

Source: U.S. Census Bureau (www.census.gov/; accessed November 17, 2010).

4

National and state flood policy mitigation in the United States

In an attempt to mitigate the mounting losses caused by floods, which total billions of dollars of damage in the U.S. every year (as outlined in the previous two chapters), federal, state, and local governments have implemented a range of regulatory and policy mechanisms. These policies range from indirect attempts to reduce flood damage through local and state-level planning initiatives to very specific and direct mitigation activities to cover and avoid losses at the federal level. As flood damages and attention to flood events have increased, so too have the attempts to mitigate losses through policy initiatives. However, the steps taken to reduce flood damages are not necessarily evenly implemented across all jurisdictions. Many direct flood mitigation policies are applied by way of a "carrot approach," where actions are not required, but instead provide some form of incentive to communities and households.

In this chapter, we present and discuss the major policy-driven flood mitigation activities and opportunities in the U.S. and specifically for our study states of Texas and Florida. We provide snapshots of the mitigation steps communities are taking through policy tools available to them at multiple jurisdictional levels. When possible, we also summarize and compare the status of flood mitigation among local communities in Texas and Florida. This chapter follows a top-down logic for describing programs and policies available for flood mitigation. First, we examine large, federal-level programs, including the NFIP and the CRS. Second, we briefly describe mitigation grant programs that are available to states and communities through federal funding mechanisms within FEMA. Finally, we discuss the state and local-level policies specific to Texas and Florida.

Federal flood mitigation

The national flood insurance program

By far the most comprehensive and widely implemented form of flood mitigation occurs through policies and programs developed and implemented through the

U.S. government's National Flood Insurance Program. The NFIP was established in 1968 to provide flood insurance to floodplain residents and businesses. Prior to 1968, the only way to assist flood victims was federal relief, which took the form of disaster loans and grants. However, the increased burden on the federal treasury caused policy makers to examine the feasibility of insurance policies against flood losses as an alternative to federal aid (Pasterick, 1998: 125–155).

The NFIP consists of three components: risk identification, floodplain management, and flood insurance. Risk identification identifies areas that are vulnerable to floods to define levels of risk and determine actuarial rates. The result of this risk analysis is the flood insurance rate map (FIRM), which contains, among other delineations, the boundaries of the regulatory 1% flood (often referred to as the "100-year flood" or base flood).

The second component, floodplain management, is considered a requirement by FEMA in order for a community to participate in the NFIP. FEMA uses the 1% or 100-year floodplain maps as a basis for mitigation action and as a minimum requirement for participation of local governments (Burby *et al.*, 1988). Localities must enforce the mitigation requirement (minimum building elevation), and compliance is accomplished through the use of a permitting program (e.g. building permits, subdivision regulations). Communities must also enact a floodplain management ordinance that meets or exceeds NFIP minimum standards. In general, for a community to participate in the NFIP, it must enact a floodplain management ordinance with the following standards:

- review and permitting of development in the 1% floodplain, or special flood hazard area (SFHA);
- new, substantially damaged, or substantially improved residential structure must be elevated above the elevation of the 1% flood; and
- new, substantially damaged, or substantially improved non-residential structures must be elevated above the elevation of the 1% flood or dry flood-proofed.

The third, and likely most recognized part of the NFIP, is its role in providing flood insurance. FEMA, through the NFIP, writes or underwrites flood insurance for participating NFIP communities in the U.S. Individuals can purchase flood insurance directly through sanctioned FEMA representatives or through a traditional private insurer in what is known as the "write your own" program. Residents in non-participating NFIP communities do not have the opportunity to purchase insurance through the NFIP.

A few characteristics make the NFIP unique to the insurance industry. First, in nearly all cases, flood insurance purchasers are held to a 30-day waiting period before the flood insurance coverage goes into effect. This essentially eliminates the ability for the insured party to purchase flood insurance when there is an imminent risk of flooding. Second, the coverage amount is capped. For example, a typical residential structure is limited in coverage to $250000 in building coverage and $100000 in personal property. While this may appear to be a large ceiling, in reality it is far less than the value of many structures, especially considering many coastal property values. Finally, there is a mandated requirement to purchase flood insurance for structures located within the SFHA that are being purchased by loan. As a result of habitual noncompliance, this requirement has been more forcefully implemented through lenders and loan servicers, requiring them to determine and document whether a structure is in the SFHA and ensure that the mortgager maintains flood insurance throughout the life of the loan.

Since community and individual participation in the NFIP was completely voluntary until 1974, the rates of purchase before that time were quite low. However, the Flood Disaster Protection Act in 1973 strengthened the NFIP so that participation of communities in NFIP became a condition of eligibility for certain types of federal assistance. The passage of this Act sparked heavy involvement in the NFIP. For example, up until 1973, about 2200 communities had joined the program voluntarily; by 1977, participation had swelled to approximately 15000 communities (FEMA, 2002). As of 2010, the NFIP had over 23000 participating communities and over 5.6 million flood insurance policies in force (FEMA, 2010). Interestingly, 38% of policies nationwide are in force in Florida, while only 12% of the 5.6 million policies cover properties in Texas (FEMA, 2010).

It has become clear that the NFIP resulted in a number of significant achievements in floodplain management, including more widespread public identification of flood hazards and reduced development in floodplains by raising the cost of land preparation, construction, and insurance (Holway and Burby, 1990; U.S. Interagency Floodplain Management Review Committee, 1994). However, for several reasons the program is not without its adverse impacts and unintended consequences. First, the NFIP includes no other requirements for land use controls in floodplains, except building elevation. The NFIP's elevation standard was indeed effective, but its overall impact is considered limited in the absence of additional land use regulations (Holway and Burby, 1993). Furthermore, mandatory insurance and elevation requirements only apply to 1% floodplain zones (as opposed to less frequent or lower probability floods) and primarily only guide new construction in those areas; its effect on existing structures is limited at best.

A second problem with, and frequent criticism of, the NFIP is out-of-date FIRMs, which, as mentioned above, are the maps used to identify the flood risk zones and provide policy guidance. In many cases, these maps are years out of date and do not reflect the true flood risk to households and communities due primarily to urban and suburban development. Recent figures indicate that 33% of the maps are more than 15 years old, and another 30% are 10–15 years old (Birkland *et al.*, 2003: 46–54).

This problem is currently being addressed through FEMA's Map Modernization Program, which aims to update maps to digital (GIS) formats for 92% of the U.S. population. Nonetheless, in order to successfully mitigate risk through insurance, a more accurate statistical and spatial depiction of this risk must be identified.

Third, one of the biggest criticisms of the NFIP centers on the population with an interest in purchasing flood insurance and the flood risk to which they are subjected. The availability of insurance in high-risk areas often leads to greater risk exposure, resulting in what is referred to as a "moral hazard." In other words, the probability of loss is increased by policy holders behaving more carelessly than they would without the availability of insurance (Kunreuther and Roth, 1998). Flood insurance is a prime example of a moral hazard, as certain individuals and businesses would likely not live in vulnerable locations without either the expectation of financial support or the availability of low-cost insurance. The former was a major impetus in the development of the NFIP; the latter reflects the situation of the NFIP today. In addition, because flood insurance is primarily only purchased by those who are vulnerable to flooding, a problem known as adverse selection, the NFIP cannot effectively pool risks. In other words, the demand for insurance is at a high rate among those who are at the most risk of flooding; a significant flood event can force insurers to pay out large sums at the same time that little or no premiums are being paid into the program. This is the primary reason that affordable flood insurance is not available through the private sector (Bagstad *et al.*, 2007).

This is especially true for repetitive loss properties – those properties that have had: (1) four or more paid flood losses of more than $1000 each; (2) two paid flood losses within 10 years that exceed the current value of the insured property; or (3) three or more paid losses that equal or exceed the current value of the insured property. In a private insurance market, properties with repetitive losses would be dropped from coverage; in the case of the NFIP, the government is required to continue coverage unless the property is substantially damaged (greater than 50% damaged). This clause essentially incentivizes continued development in flood-prone areas (Bagstad *et al.*, 2007).

The increased exposure, or moral hazard, coupled with adverse selection and repetitive losses, has led to payouts from the NFIP that consistently exceed its income. This fiscal problem was initially intended to be remedied by raising premiums and/or dropping coverage. However, a combination of payouts from large hazard events and the lack of actuarially sound rates mean FEMA is often forced to borrow money from the federal treasury to cover its deficit. Since its inception in 1968, the NFIP has borrowed over $17 billion dollars (in 2008 dollars); $16.6 billion of the cumulative amount followed hurricanes Katrina and Rita in 2005 (King, 2008). Prior to these two catastrophic flood events, the program had borrowed $225 million dollars of unreimbursed funds (King, 2008).

Finally, critics of the NFIP also cite its effect on subsidizing development in ecologically important areas. This not only places human settlements at risk from flood events, but also reduces or destroys the natural functions of floodplains in reducing floods (see Chapter 6 for more details on the effects of altering naturally occurring wetlands).

The community rating system

While participation in the NFIP is optional, communities are, in a sense, compelled to join because, without participation, homeowners would be forced to purchase insurance in the private sector, where it may not be available or financially feasible. There are thousands of communities with properties that must be insured due to mandatory flood insurance requirements, especially relative to communities with little to no flood risk. However, in 1990, FEMA introduced the CRS as a way to encourage local jurisdictions to exceed the NFIP's minimum standards for floodplain management. Participating communities adopt flood mitigation measures that are heavily weighted toward non-structural activities in exchange for an NFIP premium discount of up to 45%. The CRS program categorizes planning and management activities into four "series": public information, mapping and regulation, flood damage reduction, and flood preparedness.

Specifically, public information (Series 300) activities indicate the ability of a local jurisdiction to inform its residents about flood hazards, insurance and household protection measures. Six public information activities comprise this series: *310 Elevation Certificates*; *320 Map Information Service*; *330 Outreach Projects*; *340 Hazard Disclosure*; *350 Flood Protection Information*; and *360 Flood Protection Assistance*. Mapping and regulation (Series 400) activities involve both critical data needs and regulations that exceed NFIP minimum standards. Activities that make up Series 400 are: *410 Additional Flood Data*; *420 Open Space Preservation*; *430 Higher Regulatory Standards*; *440 Flood Data Maintenance*; and *450 Stormwater Management*. Damage reduction (Series 500) activities require specific mitigation techniques, such as acquiring, relocating, or retrofitting existing buildings. This series is composed of four activities: *510 Floodplain Management Planning*; *520 Acquisition and Relocation*; *530 Flood Protection*; and *540 Drainage System Maintenance*. Finally, flood preparedness (Series 600) entails coordinating local agencies and their programs to minimize the adverse effects of floods. Specific activities in series 600 are: *610 Flood Warning Program*; *620 Levee Safety*; and *630 Dam Safety* (for more information see: http://training.fema.gov/EMIWeb/CRS/).

Credit points are assigned for 18 different flood mitigation "activities" falling within designated series, but activities do not carry the same amount of credit. As shown in Table 4.1, more points are available to communities that implement what should be more effective flood mitigation actions. For example, under series

Table 4.1 *Nationwide summary of CRS activity points*

Activity	Maximum possible points	Average points earned	Maximum points earned	Percentage of communities credited
300 Public Information Activities				
310 Elevation Certificates	162	69	142	100
320 Map Information Service	140	138	140	95
330 Outreach Projects	380	90	290	86
340 Hazard Disclosure	81	19	81	61
350 Flood Protection Information	102	24	66	87
360 Flood Protection Assistance	71	53	71	48
400 Mapping and Regulatory Activities				
410 Additional Flood Data	1346	86	521	29
420 Open Space Preservation	900	191	734	83
430 Higher Regulatory Standards	2740	166	1041	85
440 Flood Data Maintenance	239	79	218	68
450 Storm-water Management	670	98	490	74
500 Flood Damage Reduction Activities				
510 Floodplain Management Planning	359	115	270	20
520 Acquisition and Relocation	3200	213	2084	13
530 Flood Protection	2800	93	813	6
540 Drainage System Maintenance	330	232	330	69
600 Flood Preparedness Activities				
610 Flood Warning Program	255	93	200	30
620 Levee Safety	900	198	198	1
630 Dam Safety	175	66	87	81

Source: FEMA (2007b) *National Flood Insurance Program Community Rating System Coordinator's Manual* (http://training.fema.gov/EMIWeb/CRS/2007%20CRS%20Coord%20Manual%20Entire.pdf; accessed July 1, 2010).

400 most of the available credit points (2740) are found in *Higher Regulatory Standards*, which includes activities such as requiring freeboard on structures built in floodplains, preserving natural and beneficial functions, lowering the substantial improvement threshold, and protecting the storage capacity of floodplains

Table 4.2 *CRS flood insurance premium discounts by class*

		Discount, %	
Credit points	Class	SFHA	Non-SFHA
4500+	1	45	5
4000–4499	2	40	5
3500–3999	3	35	5
3000–3499	4	30	5
2500–2999	5	25	5
2000–2499	6	20	5
1500–1999	7	15	5
1000–1499	8	10	5
500–999	9	5	5
0–499	10	0	0

Source: FEMA (2007b) *National Flood Insurance Program Community Rating System Coordinator's Manual* (http://training. fema.gov/EMIWeb/CRS/2007%20CRS%20Coord%20 Manual%20Entire.pdf; accessed July 1, 2010).

from fill and construction. In contrast, only 239 points are available for *Flood Data Maintenance*. The same imbalance can be seen in series 500, where the most points (3200) are available for acquiring and relocating insurable buildings in the floodplain. Conversely, only 359 points are available for *Floodplain Management Planning* and 330 points for *Drainage System Maintenance*. Thus, the points are generally weighted more toward non-structural activities perceived as effective.

The total number of credit points obtained by a participating locality is used to determine the extent of insurance premium discounts. Credit points are aggregated into "classes," from 9 (lowest) to 1 (highest). Communities awarded a higher CRS class will have implemented a greater number of the 18 flood mitigation measures and therefore receive a higher premium discount for insurance coverage. Discounts range from 5% (class 9) to 45% (class 1), depending on the degree to which a community plans for the adverse impacts of floods (see Table 4.2 for more detail). While the local jurisdiction takes responsibility for implementing each activity, the individual homeowner receives the discount on their national flood insurance premium. The CRS program is also revenue neutral: as premium discounts are applied to communities practicing better floodplain management, base flood insurance rates are scaled upward (Congressional Budget Office [CBO], 2009).

In 2009, there were 1110 participating CRS communities in the U.S., a small proportion of the 23 000+ NFIP communities. However, in terms of policies, CRS communities represent two-thirds of NFIP policies (CBO, 2009). As of 2009, of all the CRS communities, 217 (19.6%) were located in Florida. In contrast, only 45 CRS communities (4%) were located in Texas.

As of May 2008, CRS participating communities in Florida held over two million NFIP policies in force worth over $811 million in insurance premiums. Homeowners among 220 local jurisdictions saved approximately $132.5 million through CRS discounts. The City of North Miami earned the most points among all participating localities in Florida and maintains a class 5 rating. In contrast, Texas had 356 262 NFIP policies in 43 participating CRS communities, totaling almost $138 million in insurance premiums. Because local jurisdictions participate in the CRS, property owners saved over $9 million during that year alone. The City of Kemah received the highest total score among all CRS communities in Texas and is one of only three class 5 communities in Texas, thereby earning a 25% premium discount.

In many ways, the CRS provides points for activities that offset several of the NFIP's minimal floodplain regulation shortcomings. For example, through CRS activity 430 (*Higher Regulatory Standards*), communities are given point credits for interpreting "substantial improvement" as cumulative over time. In other words, instead of a property being flooded and damaged less than substantially and repaired repeatedly over time, the "cumulative substantial improvement" element requires that communities consider the repairs cumulatively over 25–45 years. This has the effect of identifying and eliminating flood-prone properties that might not fall under a repetitive loss category.

A second example is activity 420 (*Open Space Preservation*), under which point credits are given to communities that permanently preserve floodplains as open space. Additional credit is given if these areas are deed-restricted from development or have been restored to or retained to their natural state. This point allocation not only removes the risk of flooding from properties in the floodplain, but helps to preserve the natural environment from further development. To counter the issue of repetitive losses, the CRS not only gives credit for acquiring and relocating flood-prone and repetitive loss properties, but also requires the community to undertake specific actions depending on its repetitive loss category. For example, if a community has between one and nine repetitive loss properties, it is required to conduct an annual outreach project that directly targets those properties. If the community has 10 or more repetitive loss properties, it must conduct the annual outreach project as well as prepare a floodplain management plan or area analysis for its repetitive loss properties.

While it is also important to integrate flood mitigation into other local policy vehicles, such as comprehensive plans, there is discontinuity between policies adopted and the implementation of those policies over time (see, for example, Brody and Highfield, 2005: 159–175). The CRS is an ideal measure of the degree to which local jurisdictions adopt non-structural flood mitigation techniques because to receive a score, the FEMA program requires and confirms that activities have been implemented through field verification visits. And, while the CRS is a voluntary program, it strives through incentives to further strengthen two of the intended goals of the NFIP: risk assessment and floodplain management. However, in doing so, the CRS also acts as a counter-balance to many of the NFIP's unintended consequences. The impacts and effectiveness of the CRS program on flood losses are discussed further in Chapter 7.

It should also be noted there are other supplemental, federally derived grant programs aimed at facilitating mitigation that are less central to our study. Texas and Florida, as well as any other state in the U.S., are eligible to request grant funds through a series of federally funded programs, including: the Flood Mitigation Assistance Program (FMA), the Hazard Grant Mitigation Program (HGMP), and the Pre-Disaster Mitigation Program (PDM). There are also two programs related to repetitive losses: the Repetitive Flood Claims Program (RFC) and Severe Repetitive Loss Program (SRL).

While the above-mentioned grants are federally derived, selected state agencies are responsible for their administration. In our two states of interest, Florida programs are administered by the Florida Department of Emergency Management; in Texas, mitigation grant programs are administered by the Texas Water Development Board. Although funded primarily by the federal government, matching funds of 25% or greater are required under these programs, with the exception of RFC and SRL, which may only require 10% matching funds. Generally speaking, these grant programs are aimed at providing funding for hazard mitigation activities. Only the FMA, RFC, and SRL are specifically aimed at reducing flood losses. Mitigation for other hazards is allowed through both the HGMP and PDM.

Local planning for flood mitigation

Although far less comprehensive and more "patchy" in geographic scope, the role of local-level planning and policies in flood mitigation cannot be overlooked. In many cases, local governments play an important role in reducing the adverse impacts of floods, especially through land use planning and regulations. The idea of integrating hazard mitigation and land use planning has a long history. Gilbert White (1936) and other scholars (Burby *et al.*, 1985, 1999: 247–258; Godschalk *et al.*, 1989), for example, have long argued that losses in property and lives from natural hazards could be minimized through local land use planning initiatives. Despite these early calls, federal, state, and local governments have often overlooked the importance of not only hazard mitigation itself, but also mitigation through development management (Burby, 2005: 67–81).

In the U.S., local governments are almost exclusively responsible for land use planning and regulation (Burby *et al.*, 1997; Hoch *et al.*, 2000); thus, their essential role in hazard mitigation is critical. However, because land use planning and local-level policy mechanisms occur at the bottom of the regulatory hierarchy, they vary greatly in their existence, approach, and comprehensiveness. What emerges is an uneven and difficult to discern patchwork of local policies across regions with similar geophysical and socioeconomic vulnerabilities. For example, as mentioned in Chapter 1, in 1986, Florida established minimum criteria for local government comprehensive plans through Rule 9J-5 of the Florida Administrative Code, adopted by the Florida Department of Community Affairs (DCA). This state mandate applies to both county and municipal governments. The requirements ensure a consistent format or "checklist" approach for the establishment of local government comprehensive plans that even specifies specific content. State-required elements in a comprehensive plan include: land use, housing, infrastructure, coastal management, conservation, intergovernmental coordination, capital improvement, and transportation. Flood hazards are mainly addressed in the coastal management element, which includes plans for hurricane evacuation and high-risk area management. Within the coastal management plan element, local governments must adopt specific objectives, including:

- limitation of public expenditures that subsidize development in high-hazard coastal areas; and
- protection of human life against the effects of natural disasters.

Other plan elements, such as future land use, transportation, and conservation must also address flood management issues. For example, the future land use element of every local plan in Florida must include "an analysis of the proposed development and redevelopment of flood prone areas based upon a suitability determination from Flood Insurance Rate Maps, Flood Hazard Boundary Maps, or other most accurate information available" (9J-5.006).

Once a comprehensive plan is adopted at the local level, it must be approved by the DCA before it becomes law (Burby *et al.*, 1997). The state also requires that plans undergo evaluation and revision every seven years to adjust to changing environmental and socioeconomic conditions. This top-down, prescriptive approach to planning (although sometimes criticized for its stringency) means that every county and city in Florida has at least a minimum level of flood mitigation policies and practices.

In addition to the local planning mandate, Florida also mitigates flooding through water management districts. The state is subdivided into five water management districts: Northwest, Suwannee River, St John's River, Southwest, and South (it should be noted that the water management districts do not follow true administrative boundaries but are more closely aligned with watershed boundaries, an important management tool for floodplain management and ecosystem management (see Brody, 2008). These districts are subdivisions of the Florida Department of Environmental Protection and act as another administrative layer of water management-related activities that includes flood protection. In addition to managing and supporting water quality and water supply issues, they are also charged with flood protection both through structural measures, such as the use of levees, canals, and holding areas, and non-structural techniques, including land acquisition and preservation.

Conversely, Texas has no such state requirement for county or municipal planning. Enabling legislation instead allows "home rule" cities (those with a population over 5000) to participate in zoning and other land use planning activities, which many do. Smaller jurisdictions or county governments in Texas cannot enact zoning regulations or enforce building codes. County governments can, however, issue building permits and enforce subdivision regulations but are not able to prescribe or limit *where* the development occurs by means of land use planning or regulation through zoning. These restrictions make it difficult for localities in Texas to mitigate floods through planning and non-structural policies (see Chapter 7 for more details on non-structural mitigation approaches). Specific local flood mitigation programs in both Texas and Florida are further investigated and discussed in Chapter 10.

One notable exception within our Texas coastal study area is the formation of the Harris County Flood Control District (the District), a special-purpose entity created by the Texas Legislature in 1937 in response to severe floods. The District's jurisdictional boundaries coincide with Harris County, which includes the City of Houston, along with 22 primary watersheds across 1756 square miles (4546 km²). The objective of the District is to reduce the risk of flood damage by: (1) devising the storm-water management plans; (2) implementing the plans; and (3) maintaining drainage and flood control infrastructure. Specific plans and projects are funded by a dedicated *ad valorem* property tax (set at 3.3 cents per $100 valuation) with federal support. The current 5-year Capital Improvement Program entails more than $975 million in flood reduction projects. It is important to note that the District does not have sole jurisdiction over flood-related issues in Harris County and is one of several entities involved in flood mitigation, such as the City of Houston (for more information, see: www.hcfcd.org/index.asp).

In sum, flood mitigation in the U.S. involves a complex, often confusing web of policies and programs implemented at multiple jurisdictional levels. From top-down programs, such as the NFIP, to bottom-up planning processes within municipalities, flood mitigation presents itself as a diverse patchwork quilt of sometimes conflicting initiatives. This inconsistency is exemplified by our study states Texas and Florida, which, at the local level, take polar-opposite approaches to reducing the adverse impacts of floods (see Chapter 7 for more details). One state has a top-down, coercive mandate for flood planning at the local level; the other assigns the responsibility for flood mitigation to each municipality. These differences make for a valuable comparative analysis, and, as we show in subsequent chapters, prove to be important in influencing the extent of flood losses over time.

Part II
Planning decisions and flood attenuation

5

Identifying the factors influencing flooding and flood damage

An important step in reducing the adverse effects of floods is first to identify the factors driving the degree of impact on human communities. Understanding the levers exacerbating or minimizing floods can help decision makers foster the development of more resilient communities. Because many of these levers are human induced (rather than purely natural acts), they can be manipulated, adjusted, or pressed on to mitigate potential damages and loss of life. The key to building flood-resistant communities, then, may be recognizing which, among multiple levers, exert the most force on the problem. The ones that can be altered through thoughtful policies and plans should be addressed first.

Some disciplines identify precipitation, slope, or stream density as the major causes of flooding; others point to population growth or land use policies as the basis for predicting flood events. In reality, the size and impact of floods is most likely a complex mix of all these characteristics and more. We assert that the only way to thoroughly understand the factors influencing flooding and flood damage is to break through single disciplines and assume a trans-disciplinary approach to modeling the problem. An interdisciplinary conceptual and quantitative model must include geophysical as well as socioeconomic and human built-environment characteristics. This approach will not only improve prediction, but represent a more holistic, and perhaps more realistic, representation of the actual problem. The end result is more precise information for those responsible for shaping coastal communities in the future.

In the following section, we describe the major variables contributing to flooding, flood damage, and human casualties from flood events (Table 5.1). These variables are grouped into five categories or dimensions of flood prediction: natural, built, organizational, socioeconomic, and mitigation. In subsequent chapters, we empirically test the effects of these factors on flood impacts in Texas and Florida.

Table 5.1 *Factors influencing flooding and flood damage*

Natural environment	Built environment	Socioeconomic	Flood mitigation	Organizational capacity
Basin area	Impervious surfaces	Housing values	Structural	Collaboration
Basin shape	Wetland alterations	Education	Non-structural	Competency
Topography	Development density	Population change		Individual characteristics
Precipitation	Housing units	Income		
Soils				

Natural environment characteristics

Basin area

The earliest set of conditions studied for their impacts on flooding were geophysical and other abiotic components of natural landscapes. Primarily the domains of hydrological and engineering sciences, these variables are often used in simulation modeling and single-basin case study analyses. The oldest and most commonly measured of these variables is the drainage basin or watershed area. Drainage area is consistently found to be a significant factor affecting discharge, where larger areas correspond with increased flooding potential. In fact, drainage area is such a commonly used variable in hydrological sciences, it is often used to predict streamflow characteristics, particularly for sites that lack gauges (U.S. Geological Service [USGS], 1997).

Basin shape

The shape of a drainage basin is another important natural environment variable affecting hydrological conditions. In general, basin shape influences stream peak-flow rates (Saxton and Shiau, 1990: 55–80) by determining the temporal concentration of water runoff (Matthai, 1990: 97–120). Streams in longer, narrower basins will typically peak and begin receding in the lower areas of the basin before they are affected by flows from upstream areas. In contrast, streams in more regularly shaped basins will typically have the same times of concentration, causing a faster rise to and recession from peak discharges (Matthai, 1990: 97–120). Basin shape measurements include length-to-width ratio or the "shape factor" (Horton, 1932: 350–361), circularity ratio (Miller, 1953), and elongation ratio (Schumm, 1956: 597–646). For example, the elongation ratio is calculated by dividing the diameter of a circle with the same area as that of the basin by the basin length. A large elongation ratio value is an indicator of a more regularly shaped basin, whereas a small elongation ratio is indicative of a longer, narrower basin. A more regularly shaped basin is more flood-prone because water moves more slowly out of the area following a rainfall event.

Topography

Another characteristic of basin geomorphology contributing to flooding is topography. Specifically, the slope of a watershed affects both the temporal concentration and the amount of water storage. Slopes may act in concert with or against the effects of basin shape. Generally, steeper slopes increase rainfall concentration, causing faster and higher stream peaks as well as mean annual flows (Matthai, 1990: 97–120; Stuckey, 2006). Under these conditions, water bodies tend to overflow their banks more quickly and with less warning than do more gently sloped watersheds. On the other hand, there is less depressional pooling of water on steep upper slopes where runoff sheds more quickly. Hydrologists measure topography in several ways when explaining streamflow magnitudes, including mean basin slope, basin relief, relief ratio, and mean stream slope.

Precipitation

Aside from geophysical characteristics of the landscape, climate, particularly precipitation, is perhaps the most powerful predictor of flooding. Precipitation is the primary driver of the hydrological conditions leading to flooding and associated impacts on human communities. Generally, the more rainfall, the greater the likelihood streams and rivers will crest their banks due to excessive runoff. Four characteristics of a precipitation event contribute to its flood potential: intensity, depth, duration, and distribution over a drainage basin. Precipitation depth (amount) and duration are described as the attributes of the storm exterior. The temporal and spatial distribution of precipitation is referred to as the storm interior (Bras, 1990). These concepts have important repercussions on measurement and estimation. For example, historic records of precipitation are typically collected at point locations. Yet, the amount of precipitation over an entire watershed is typically the necessary input in any hydrological study. The estimation of rainfall over areal units can be derived in several ways including the arithmetic mean of stations, thiessen polygons, isohyets, radar-based estimates, and more advanced forms of spatial interpolation such as inverse-distance weighting, splines, and kriging (Running and Thornton, 1996).

Soils

A final landscape characteristic essential for explaining the magnitude of flooding is soil. Generally speaking, soils serve three primary functions: to absorb; to store; and to release water. The amount of water that any given soil will infiltrate and retain depends primarily upon its texture and current moisture condition (Saxton and Shiau, 1990: 55–80). Numerous measures are available to quantify soil characteristics across basins. Common measures include soil permeability, holding capacity, soil thickness, and specific hydrologic grouping. Soil permeability, the ability of water to flow through a soil, is the preferred measure for our research. The potential for higher peak and mean annual flows from basins with low soil permeability is greater than that for basins with higher permeability soils, as higher permeability allows greater infiltration, more storage, and less runoff (Rasmussen and Perry, 2000). Therefore, we would expect low-permeability soils to be more prone to flooding and less desirable for building permanent structures.

Built-environment characteristics

For the most part, the natural environment characteristics described above are difficult if not impossible to change at the watershed level (although altering topography, elevation, and soils is commonly done in localized areas). Thus, they may be powerful levers on the problem of flooding, but may not be practical to move from a policy-making perspective. As already discussed, the real opportunity for reducing the adverse consequences of floods lies in the way humans build upon the physical landscape. Since flood disasters are a human-induced phenomenon, changing the way we shape our communities and development patterns is the most effective way to mitigate repetitive and costly flood events.

Impervious surfaces

A consequence of coastal development and the urbanization of landscapes is the increase in impervious surfaces (Schuster *et al.*, 2005: 263–275). Conversion of agricultural and forest lands, wetlands, and open space to urban areas can compromise a hydrological system's ability to absorb, store, and slowly release water. The result of widespread hardened surfaces is often increased flood intensity (Carter, 1961; Tourbier and Westmacott, 1981). Greater areas of impervious surface coverage correspond to a decrease in rainfall infiltration and an increase in surface runoff (Paul and Meyer, 2001: 333–365). According to Arnold and Gibbons (1996), as the percentage of impervious surfaces within a drainage basin increases 10–20%, on average, storm-water runoff nearly doubles. More recently, White and

Greer (2006) found that as urbanization in the Peñasquitos Creek watershed in southern California grew from 9% to 37%, total runoff was amplified by an average of 4% per year. When extended over the authors' study period of 1973 to 2000, this yearly runoff estimate amounts to a 200% increase. A higher level of runoff is important because it can translate into increased frequency and severity of flooding in rivers and streams.

Impervious surfaces have also been associated with increased peak discharges (Brezonik and Stadelmann, 2002: 1743–1757; Burges *et al.*, 1998: 86–97; Leopold, 1994). Under compromised hydrological conditions, the lag time between the center of precipitation volume and runoff volume is compressed so that floods peak more rapidly (Hirsch *et al.*, 1990: 329–359). This reduced lag time occurs because runoff reaches water bodies more quickly when rainfall is unable to infiltrate into the soil (Hey, 2002: 89–99; Hsu *et al.*, 2000: 21–37). For example, Rose and Peters (2001) measured peak discharge increases of approximately 80% in urban catchments with more than 50% impervious area. Similarly, flood discharge was at least 250% higher in urban compared with forested catchments in Texas and New York after similar storms (Espey *et al.*, 1965; Paul and Meyer, 2001: 333–365; Seaburn, 1969: 14). Burns *et al.* (2005) also examined mean peak discharges for 27 storms in the Croton River Basin in New York. They observed a 300% increase in a catchment with an impervious area of only about 11%. In general, there is a growing body of evidence to support the notion that urbanization increases not only runoff volume, but also peak discharges and associated flood magnitudes.

From a flooding standpoint, the concern is not the single shopping-mall parking lot, but rather the cumulative effect of thousands of individual development decisions that result in a landscape dominated by impervious surfaces. As of the last U.S. census, almost 80% of the population was living in urbanized areas. According to a recent study, between 1982 and 1997, there was a 34% increase in the amount of land in urban or built-up uses. This area of mostly impervious surface is projected to increase by almost 80% in the next 25 years, raising the proportion of the total U.S. urban land base from 5.2% to 9.2% (Alig *et al.*, 2004: 219–234). Nowhere is this urban expansion more noticeable than in the Houston, TX area, which has quickly become one of the largest expanses of impervious surfaces in the country, with a nearly uninterrupted swath of pavement approximately 60 miles (96.54 km) long and 40 miles (64.36 km) wide. Freeways, parking lots, rooftops, and urban parklands are ubiquitous across the Houston Metropolitan Area. If water is unable to drain slowly into the soil or nearby water bodies, it has nowhere to go but into people's homes and businesses. For example, as described earlier, a summer rain shower of 4 inches (10.16 cm) in Houston is enough to flood major roadways. Water pools onto road surfaces (particularly highway underpasses), which are usually the lowest-lying areas in the city, trapping motorists during intense rainstorms.

Wetland alterations

The relationship between urban development and flooding depends not only on the regional extent of impervious surfaces, but the specific location of the development within the hydrological system. Thus, the effect of development on flooding is not based solely on land-use intensity alone, but also location-based attributes. One key attribute within hydrological landscapes is naturally occurring wetlands, which are believed to provide natural flood mitigation by maintaining a properly functioning water cycle (Mitch and Gosselink, 2000; Lewis, 2001). Both anecdotal and empirical research suggests that wetlands may reduce or slow flooding. In the most comprehensive literature review to date, Bullock and Acreman (2003: 366) note that 23 of 28 studies on wetlands and flooding found that "floodplain wetlands reduce or delay floods."

Initial research on the role of wetlands in reducing flooding examined the differences between drained and natural wetlands. These studies showed that nondrained peat bogs reduce low-return period flood flow and overall storm flows when compared with drained counterparts (Daniel, 1981: 69–108; Heikuranen, 1976: 76–86; Verry and Boelter, 1978: 389–402). For example, Novitski (1979) examined four different types of wetlands and found that each had a statistically negative effect on flood flows. Later, Novitski (1985) discovered that basins with as little as 5% lake and wetlands area may result in 40–60% lower flood peaks.

Research based on simulation models also suggests that wetlands have the natural potential for reducing floods. For example, Ammon *et al.* (1981) modeled the effects of wetlands on water quantity for the Chandler Slough Marsh in South Florida. Results indicate that flood peak attenuation was greater with larger areas of marsh. The authors concluded that Chandler Slough Marsh increases storm water detention times, facilitates runoff into subsurface regimes, and is fairly effective as a water quantity control device. Ogawa and Male (1986) also analyzed a simulation model to evaluate the protection of wetlands as a flood mitigation strategy. Based on four scenarios of downstream wetland encroachment, ranging from 25% to 100% alteration, these researchers found that increased encroachment resulted in statistically significant increases in stream peak flow.

Another form of research based on direct observation, rather than simulation, also supports the idea that naturally occurring wetlands can reduce flooding events. For example, a constructed wetland experiment along the Des Plaines River in Illinois found that a marsh of 5.7 acres (2.3 hectares) could retain the natural runoff of a 410-acre (166-hectare) watershed. The same study estimated that only 13 million acres (5.26 million hectares) of wetlands (3% of the upper Mississippi watershed) would have been needed to prevent the catastrophic flood of 1993 (Godschalk *et al.*, 1999). Another empirical research method involves measuring streamflow data from stream gauge stations. Using this approach, Johnston *et al.* (1990) found that even small wetland losses in watersheds could significantly affect flooding over time.

While the body of research on wetlands as natural flood attenuation devices continues to expand, the subject is grossly understudied and undervalued in the field of environmental management. To date, no empirical studies have been conducted longitudinally, over large spatial scales, while controlling for multiple geophysical, built environment, and socioeconomic variables. Given its importance, we will return to this issue in Chapter 6 where we examine the role of wetland alteration and flooding using our own data to further shed light on the potential of wetlands as a cornerstone of a local flood mitigation program.

Development density

Land use change contributing to increased vulnerability to flooding is not only a function of the intensity or location of development, but also its regional pattern. Sprawling development patterns, typified by low-density, residential dwelling units spreading outward from urban cores, dominate much of the American landscape (Beatley and Manning, 1997; Burchell *et al.*, 1998) and are particularly prevalent in Texas and Florida. While the environmental consequences of sprawl are well studied (Arnold and Gibbons, 1996: 243–258; Benfield *et al.*, 1999; Brody *et al.*, 2006a: 294–310; Hirschhorn, 2001: 1–8; Kahn, 2000: 569–586; Kenworthy and Laube, 1999: 691–723;), the impacts of sprawl on flooding have been largely overlooked.

One result of this built-environment pattern is the over-consumption of land originally designated for other purposes. For example, South Florida has among the highest percentage change in urbanized land in the country. In Texas, from 1982 to 1997, over 1.7 million acres (688 000 hectares) of agricultural land were converted to development, more than any other state in the nation (U.S. Department of Agriculture [USDA], 2000). The Houston–Galveston area is perhaps the best example of land consumption from urban and suburban sprawl in Texas. Between 1970 and 1990 alone, Houston urbanized approximately 640 square miles (1657.6 km^2) of land, second only to Atlanta, Georgia, during the same time period (U.S. Census Bureau). Land conversion is highly correlated with impervious surfaces and the alteration of hydrological systems, as explained above. When situated in flat, low-lying areas containing naturally occurring wetlands with intense periods of precipitation, this development pattern provides a recipe for flooding and its associated adverse impacts.

Perhaps the best way to measure and observe sprawl in the context of flood vulnerability is through the concept of density. The more spatially concentrated an area of development is, the more damage a localized flood could inflict. From a structural perspective, locations with high value per square acre (hectare) have more land-based capital at risk from floods. Likewise, a flood event could adversely affect more people in areas with high population densities. On the other hand, low-density development patterns place more structures and residents at risk from flooding over a larger area, thereby increasing the overall level of community vulnerability. Low-density development also generates larger areas of impervious surfaces that increase surface runoff and can exacerbate flooding.

The key to building resilient communities is to prevent sprawl from occurring in regions vulnerable to floods. Unfortunately, our study states are poor examples of smart growth. For example, only 13% of South Florida's office space is located in its central business district (CBD), compared with a median of nearly 30% for all 13 markets in the U.S. Furthermore, from 1987 to 2002, Miami's non-CBD market grew over 60% to include nearly 30 million square feet (2 787 091 m^2) of office space. In contrast, office space in Miami's CBD increased just 4.7% over this same time period (Lang, 2003). Houston, Texas, consistently has one of the lowest populations per square mile (km) of any city in the U.S. From 1991 to 2003, 78% of wetland alteration permits were issued outside of urban areas, reflecting sprawling growth patterns associated with coastal development (Brody *et al.*, 2008a: 107–116). As a result, the Houston–Galveston area has quickly become one of the largest connected expanses of impervious surfaces in the nation. Impervious surfaces impede the ability of runoff to percolate into the soil, exacerbating the potential of bank overflows and associated flooding. It is no wonder Florida and Texas record some of the most flood-related damage among all states every year.

Socioeconomic characteristics

Flooding and associated adverse impacts do not stem solely from the amount of precipitation, the shape of a watershed, or the locations of parking lots. This natural hazard is as much about people and household composition as anything else pinpointed in the literature. Systematic research on the factors influencing floods generally overlooks socioeconomic and demographic characteristics of local communities (see Peacock *et al.*, 1997). However, these may be significant factors in predicting the likelihood and extent of flood disasters and should at least be incorporated as control variables in hydrological or engineering models.

Housing value and income

Principally, the degree of wealth in a community frequently relates to the impact of a flood. Wealthier communities often have the financial capacity, both at the budgetary and household levels, to effectively mitigate flooding through various structural and non-structural techniques. At the same time, however, these communities have greater financial capital (e.g., more expensive houses) that could be lost to damaging floods.

Education

Education is another important characteristic that may affect the extent of local flooding. An educated and aware public is more likely to make informed decision as to where to live, as well as support local mitigation activities that can reduce future negative impacts. Education typically refers to the number of years of formal education an individual has completed. This measure consistently correlates with greater awareness of flood issues and adoption of household mitigation strategies to reduce the impacts of potential floods. Education is also a community-wide phenomenon initiated by a local jurisdiction. Local governments can play an important role in raising awareness of flood threats and appropriate responses through technical assistance, media outlets, written materials, and workshops.

Population change

Population growth patterns based on economic development comprise another issue important to understanding coastal flooding. With over 50% of the U.S. population residing in coastal areas, local decision makers are finding it increasingly difficult to facilitate the development of flood-resilient communities. Rapid urban and suburban development in the coastal zone has placed more people in areas susceptible to flooding. These problems are exacerbated within major population centers, particularly the Houston–Galveston and Miami–Dade metropolitan statistical areas, where population growth, sprawling development patterns, and the alteration of hydrological systems has created some of the most vulnerable communities in the nation. Population trends are based on complex interactions among broad economic signals, land values, aesthetics, personal lifestyle choices, and public planning decisions. These forces should be considered when modeling flood outcomes because they all contribute to the extent and impact of flood events.

The variables above represent just some of the socioeconomic and demographic characteristics that will inevitably influence flooding and associated damages. Other factors that should not be overlooked when predicting the impacts of floods include, among others, age, race, housing tenure, family size, gender, and residential ownership.

Table 5.2 *Flood mitigation strategies and techniques*

Structural strategies	Non-structural strategies
Retention	Stand-alone flood plans
Channelization	Setbacks and buffers
Debris clearing	Land acquisition
Levees	Zoning and land use restrictions
Dams	Protected areas
	Education
	Intergovernmental agreements
	Computer models/forecasting
	Specific polices in a comprehensive plan
	Training/technical assistance
	Referendums
	Community block grants
	Land development codes
	Construction codes

Flood mitigation techniques

Perhaps the strongest and most movable lever for reducing the negative effects of floods is mitigation. The way in which local governments plan for and respond to the threats of chronic flooding may offer the greatest opportunity to limit impacts on property and human safety. However, a wide array of policies and strategies are available that local decision makers can choose to adopt (Table 5.2). Each technique has its own potential for mitigating the rising costs of floods and fostering sustainable long-term local economies.

Local flood mitigation techniques are usually separated into two major categories: structural and non-structural (Thampapillai and Musgrave, 1985: 411–424). Structural approaches involve building and construction projects to actively secure human settlements. These techniques tend to most visibly alter the existing landscape, and include seawalls, levees, dams, channels, and revetments. Structural approaches to flood management also usually involve large amounts of financial capital, long timeframes, and can impose negative impacts on the natural environment. In contrast, non-structural techniques for flood mitigation are most often based on policies that guide development away from vulnerable areas, such as floodplains or river bottoms (Alexander, 1993). This approach includes implementation of both regulatory and incentive-based policies in an effort to shape development patterns that are more resilient to flooding over the long term. In many instances, the most effective flood management programs utilize a mixture of structural and non-structural mitigation techniques tailored to a locality's specific contextual conditions.

Structural approaches

The initial efforts at flood mitigation in the U.S. focused mostly on large-scale structural techniques, such as those implemented after the Mississippi River flood in 1927 (Birkland *et al.*, 2003: 46–54). The Flood Control Act of 1930 subsequently dedicated funds to build structural flood control works, such as levees, floodwalls, and fills, many of which are still standing. Other structural mitigation approaches that actively alter the physical landscape include the use of channel and land-phase structures to control floods. Channel-phase structures include dykes, dams, reservoirs, reducing bed roughness, and altering stream channels. Land-phase structural methods take place outside of a channel and include modified cropping practices, erosion control, re-vegetation, and slope stabilization (Alexander, 1993).

In the 1950s, researchers and public decision makers began to realize the limitations of structural approaches to flood management. First, when flood events exceed the capacity of a flood control structure, the resulting flood damages are significantly higher than if the area had been unprotected and thus less populated (Burby *et al.*, 1985; Larson and Pasencia, 2001: 167–181; Stein *et al.*, 2000; White, 1945; White *et al.*, 1975). Second, structures such as levees can raise the normal level of a river and increase the velocity of water pulsing downstream. By constricting a waterway and hardening its banks, these structures increase the probability of downstream flooding (Birkland *et al.*, 2003: 46–54). Third, structural approaches to flood mitigation, such as dams, can bring a false sense of security to residents living downstream (Burby and Dalton, 1994: 229–238). The perception that areas protected by dams are completely safe can encourage new development, increasing the risk of human casualties or property damage if the structure either underperforms or is breached during a storm event (Burby *et al.*, 1985). In 2005, Hurricane Katrina breached the levees protecting New Orleans, Louisiana, and flooded residential developments built on what was thought to be a safe area. Katrina quickly became the costliest and one of the deadliest storms in the history of the U.S.

Fourth, structural mitigation measures are usually extremely costly. Since the 1940s, the U.S. Army Corps of Engineers (USACE) has spent over $100 billion (in 1999 dollars) on structural flood control projects (Stein *et al.*, 2000). While non-structural alternatives may provide equal benefits at lower cost, structural approaches to flood mitigation have been shown to reduce the adverse impacts of floods. For example, according to USACE, although flood damages from 1991 to 2000 totaled approximately $45 billion, structural flood control measures averted an additional $208 billion of damage (USACE, 2002). Even though structural solutions to flood control may save money over the long term, their initial costs are usually very high. A final drawback of structural mitigation measures is that these structures often cause irreversible negative environmental impacts, such as the alteration of naturally occurring wetlands, degradation of fish and wildlife habitats, reduced water quality, and compromised function of hydrological systems (Abell, 1999).

Non-structural approaches

Non-structural approaches to flood mitigation have been advocated by planning researchers for some time (see, for example Burby *et al.*, 1997; Godschalk *et al.*, 1999), but are only recently gaining widespread acceptance at the local level due to their effectiveness and reduced financial burden. There exists a wide spectrum of non-structural techniques, including land use planning and zoning tools, education and training programs, environmentally sensitive area protection, flood forecasting and warning models, and insurance incentives. Many of these non-structural flood mitigation strategies come from the NFIP, which was established in 1968 as a response to mounting flood losses across the U.S. The NFIP has, by many accounts, successfully integrated flood mitigation into the regulatory fabric of many local communities. However, the program is not without deficiencies. For years, critics have raised concerns about the NFIP's underlying goal of subsidizing and thus encouraging development within floodplains, the overall equitability of the program, and the escalating financial burden of repetitive losses (Platt, 1999). One of the program's most notable shortcomings is that it allows developers to fill or alter floodplains to raise the floor elevations of structures in the 100-year floodplain (Birkland *et al.*, 2003: 46–54). Although this allowance may serve as a protective step for residential and commercial developments in areas vulnerable to flooding, it may increase the risk of flooding in adjacent and downstream areas.

Given a favorable political setting, non-structural flood mitigation may be most easily achieved through spatially targeted planning policies. Several scholars, starting with Gilbert White as far back as 1936, have argued that local land use planning techniques can facilitate the development of communities resilient to the adverse consequences of flooding (Burby *et al.*, 1985, 1999: 247–258; Godschalk *et al.*, 1989). This growing body of research argues that the public sector has overlooked the importance of not only hazard mitigation itself, but also mitigation through local-level planning and development management (Burby, 2005: 67–81). In this sense, incorporating mitigation techniques into local comprehensive planning may be the greatest opportunity for reducing the adverse effects of floods.

Place-based land use regulations, such as use restrictions, clustering, conservation overlay zones, and transfer of development rights, can work to avoid flood losses by directing growth away from vulnerable or sensitive areas. For example, in Portland, Oregon, over 162 acres of flooded properties have been purchased since 1997 (ASFPM, 2004). These purchases are complemented by stringent land use controls, including restrictions on all residential development in flood hazard areas and the use of environmental overlay zones to protect natural features such as wetlands and riparian areas that help reduce flood events as well as flood damages. Proactive planning measures that focus development either outside of the 100-year floodplain or away from flood-prone water courses can minimize flood damages, while at the same time protecting critical natural habitats and maintaining the integrity of key hydrological systems (Whipple, 1998).

Other non-structural approaches to flood mitigation that may complement traditional land use policies, such as zoning and subdivision ordinances, include public education and training, taxation and fiscal incentives, land acquisition, flood warning systems, and directing public infrastructure investments where building is most appropriate. While these and other techniques show promise, a lack of demonstrated effectiveness has prevented them from being fully embraced by local decision makers (Burby *et al.*, 1985; Olshansky and Kartez, 1998).

Flood mitigation strategies can be integrated into local land use planning initiatives either as a stand-alone flood plan or as part of a comprehensive plan. A comprehensive plan serves as an overall blueprint for community development by inventorying existing conditions, setting goals for desired future development patterns, and crafting actions to achieve them (Nelson and French, 2002: 194–207). While a single plan is more targeted, a comprehensive plan reaches more areas of community development and will affect a larger number of citizens. Through this approach, flood policies can be incorporated into related issues, such as land use, housing, public infrastructure, economic development, and transportation. By piggybacking off these plan elements, flood mitigation techniques can be effectively more assimilated into the fabric of community development (Burby *et al.*, 1999: 247–258).

Multiple studies have demonstrated the effectiveness of incorporating flood mitigation techniques into comprehensive plans (Brody, 2003b: 191–201; Burby *et al.*, 1997, Godschalk *et al.*, 1999). The general consensus is that communities with plans are better prepared for flood disasters than those without them. State mandates requiring localities to adopt comprehensive plans have high-quality hazard mitigation components and facilitate more resilient forms of community development (see Berke and French, 1994: 237–250, Burby and Dalton, 1994: 229–238, Burby *et al.*, 1997, among others). Only recently have researchers begun to consider the outcomes of strong planning with regard to reduced flood impacts. For example, Burby, (2005: 67–81) estimated that if local plans had been mandated across the U.S., insured flood losses to residential property from 1994 to 2000 could have been reduced by 0.52% and by an additional 0.47% if states had required consideration of natural hazards in local plans. This reduction translates into approximately a $200 million savings (in year 2000 constant dollars) during the study period.

An increasing number of states now require hazard mitigation in their compre-hensive plans. Currently, 11 states mandate that local governments have compre-hensive plans with hazard mitigation elements. Among them, four states (Florida, California, North Carolina, and Oregon) require a consistent format for local plans and specify their contents. Overall, 16 states mandate local comprehensive plans, but the hazard mitigation element is optional (Kang, 2009).

If land use policies have been advocated for decades as a key component of local flood mitigation programs, why are they not ubiquitous in codes across the U.S.? For one, planners are preoccupied with more immediate problems, such as housing, unemployment, and crime. Floods and other natural hazards are less of a priority because they have a low probability of occurrence and are seemingly uncontrollable events driven more by fate than by policy (Mileti, 1999). Second, the up-front financial costs for mitigating floods are high, but the benefits are dif-ficult to detect. It takes a long time to observe the positive effects of non-struc-tural techniques; elected officials who want to demonstrate immediate results to their constituents might be reluctant to adopt these policies (Berke and French, 1994: 237–250). Third, local governments may shy away from adopting stringent land use codes for fear of future legal objections and a potential backlash from voters with a pro stance on private property rights (Platt, 1999). Finally, land use policies are confined to single jurisdictions that typically do not adhere to natural boundaries. This fragmented pattern of local land use control makes it difficult to address issues that occur at floodplain, watershed, or ecosystem scales (Birkland *et al.*, 2003: 46–54; Szaro *et al.*, 1998: 1–7).

Despite their limitations, proactive policies and strategies are the cornerstone for developing flood-resilient coastal communities, because such policies ultim-ately direct how we interact with the physical landscape. However, there have been few large-scale studies on the effectiveness of various flood mitigation techniques. How do we know which strategies are more likely to reduce loss or save lives than others? Is there a combination of activities that work synergistically to reduce flood loss? Addressing these and related questions through an empirical research approach can provide invaluable guidance to local decision makers on what works best to mitigate adverse impacts from floods.

Organizational capacity

The reduction or avoidance of losses from floods rests not only on the mitiga-tion strategy, but also the strength of the organization adopting it. Government organizations with greater planning capacity will be more likely to implement appropriate flood reduction measures. Capable, resource-rich organizations, such as planning or emergency management departments, will have greater success

---

Box 5.1 Characteristics of organizational capacity

- Commitment
- Sharing information
- Sharing resources
- Leadership
- Available staff
- Public participation
- Long-range planning
- Hire and retain staff
- Public officials
- Verbal communication
- Networks
- Financial resources
- Data quality
- Adjustable policies
- Human ecology

---

at dealing with chronic flooding than those that are underfunded or receive little attention.

The term "capacity" is usually measured as the number of planning staff members devoted to drafting a local plan. However, this overly narrow interpretation fails to capture many characteristics that facilitate public entities in adopting and implementing mitigation strategies. Here, we conduct a more inclusive analysis of organizational capacity by conceptualizing the term as the ability to anticipate flooding, make informed decisions about mitigation, and implement effective policies (Honadle, 1981: 575–580). Key characteristics of organizational capacity include adequate financial resources, staffing, technical expertise, communication and information sharing, strong leadership, and a commitment to flood protection (Grindle and Hilderbrand, 1995: 441–463; Handmer, 1996: 189–197; Hartig *et al.*, 1995: 1–10; Hartvelt and Okun, 1991: 176–183;). This broader understanding of capacity is not based solely on the amount of funding or technical expertise, but also on the capability of individuals within an organizational unit to work together to attain a common goal (Box 5.1).

Organizational capacity thus constitutes a foundation of human resources on which flood mitigation programs can be built. For example, past studies have found that a more dedicated planning staff and financial resources lead to higher-quality mitigation policies (Burby and May, 1998: 95–110). Stronger planning agency capacity translates into greater technical expertise and number of personnel who can be devoted to implementing flood mitigation techniques (Brody, 2003c: 733–754;

Laurian *et al.*, 2004: 555–577; Olshansky and Kartez, 1998). Also, greater financial resources can lead to more extensive engineering approaches to mitigation or more thorough community-wide programs that work to prepare residents for flooding events.

The overall level of organizational commitment is another critical factor contributing to a strong local flood management program and the potential avoidance of losses. A local government may have the financial backing to construct a flood mitigation program, but lack of commitment from both staff and elected officials could result in a failure to implement the necessary policies (Handmer, 1996: 189–197; Ivey *et al.*, 2002: 311–331). Multiple studies (Berke *et al.*, 1996: 79–96; Brody, 2003b: 191–201; Burby *et al.*, 1997; Dalton and Burby, 1994: 444–461) have noted that the degree of local governmental commitment associated with natural hazards, such as floods, is a key characteristic in the implementation of mitigation strategies. Strong organizational commitment to protecting residents from floods should lead to the implementation of more extensive flood mitigation strategies because agencies will emphasize the importance of reducing the adverse impacts of floods during planning processes.

Another important indicator of local organizational capacity for reducing flood impacts is the ability to adjust policies in response to a flood-related problem. Planners and floodplain administrators must be flexible in their decisions so they can accommodate changing conditions of the built and natural environment, sudden shifts in local interests and politics, and a steady stream of new and often conflicting information. Hazard mitigation plans and strategies thus need to be geared toward uncertainty and surprise, with reasoned expectations about how existing conditions will respond to management actions (Holling, 1996: 733–735). For example, development restrictions in flood-prone areas can be conceived in an experimental manner. If the policy succeeds in meeting its objectives of reducing flood losses, then expectations are affirmed and residents are protected. If the policy fails, an adaptive design still allows for learning so that subsequent decisions will be made based on increased understanding. Overall, adaptive approaches to flood management ensure that organizations responsible for implementing specific actions are responsive to variations in ecological and human systems and are able to react quickly with effective management tools and techniques (Handmer, 1996: 189–197; Westley, 1995: 391–427).

It is important to note that local public organizations cannot operate in isolation, but must interface with a larger community composed of stakeholder networks, complex relationships, and a mix of human values (Brody, 2008). At the local level, flood mitigation policies are usually adopted and implemented through a collaborative process involving multiple actors with varying opinions. These groups include other government entities, as well as private and nongovernment interests.

Stakeholder groups and individuals bring to the decision-making table valuable knowledge and innovative ideas about the community that can improve the quality of plans and policies. It is often argued that collaboration can act as a powerful lever for generating trust, credibility, and commitment to the implementation of policies (Innes, 1996: 460–472; Wondolleck and Yaffee, 2000) and flood planning should be no different. Collaborative activities within and across organizations include sharing of data and information, communication, establishment of informal networks, and joint project management (Ivey *et al.*, 2002: 311–331).

The overall measure of organizational capacity can be broken down into several sub-variables, each with its own potential effect on local mitigation and corresponding flood damage. *Collaboration-based* variables include a strong line of communication, sharing information, and pooling of resources across organizational units. *Competency* variables pertain to the number of staff, level of financial resources, quality of data, and the ability to retain key personnel over the long term. Finally, the *individual characteristics* component applies to a personal commitment to flood mitigation, strong leadership within an organization, the ability to think and act over long time frames, and the aptitude to see the interplay between human and natural systems. Each one of these sub-components of organizational capacity can have its own effect on the degree of local flooding impacts.

6

The role of wetlands: federal policies, losses, and floods

As outlined in Chapter 5, the role of naturally occurring wetlands in regulating streamflow and reducing floods is critically important, especially in the low-lying areas of Texas and Florida. Despite this recognized function, regulations to limit the loss of naturally occurring wetlands have been a moving target since their inception. From the primary agency charged with permitting their alterations, to a patchwork of state and local policies, wetland regulation in the U.S. is a prime example of a constantly evolving environmental policy.

In studying this changing policy climate, previous research has pointed to difficulties mitigating wetland loss through the federal permitting process and has increasingly linked wetland alteration with flooding. This chapter addresses the importance of naturally occurring wetlands by linking federal policy administered by the USACE to record wetland loss and its regional effects on streamflow, flooding, and flood damage. First, we describe the progression of federal wetland policy and the permitting procedures of the USACE. Second, we evaluate the types of wetland loss in Texas and Florida as a result of these permitting procedures. Finally, we analyze the effects of wetland permitting and loss on flooding and flood damage. Our analysis shows that permits issued to alter naturally occurring wetlands in Texas and Florida increase both the degree and impact of flood events.

Federal wetland policy in the U.S.: Section 404 of the Clean Water Act

Federal wetlands protection, at least in some capacity, began with the passage of the Federal Water Pollution Control Act of 1972, also known as the Clean Water Act (CWA). This act initially included no references to wetlands and was primarily geared towards wastewater treatment and disposal. However, in debating the CWA, Congress recognized that the protection of water quality must reach beyond point sources. In both the House and Senate bills that were debated during the regulatory overhaul of the 1972 Act, the term "navigable waters" was openly intended to take on the broadest possible interpretation allowed by the U.S. Constitution (HR REP. NO. 92–911, 1972).

To Congress, the USACE seemed the obvious selection to implement a permitting program dealing with "navigable waters," especially considering their historical permitting experience with the Rivers and Harbors Act. However, the CWA was geared almost exclusively toward pollutant reduction, thus additional oversight from the U.S. Environmental Protection Agency (EPA) was provided. As a result, the USACE administers the Section 404 permit program, and the EPA controls the substantive water quality protection criteria that Section 404 permit applicants must meet (Downing *et al.*, 2003: 475–493). The EPA also has the authority to veto USACE permit decisions, although this power is seldom used. For example, from 1972 to 1990, the USACE issued roughly 10 000 permits per year; the EPA vetoed only 11 projects during this time period (Steiner *et al.*, 1994: 183–201). More recent figures from the EPA show that approximately 80 000 Section 404 permits are issued every year in the U.S. but only 12 permits have been vetoed since 1972. Of these 12 vetoes, two occurred in Florida and to date, no Section 404 permit has ever been vetoed in Texas.

The definition of "navigable waters" has been and continues to be the lynchpin for federal wetlands protection and permitting decisions under Section 404. Following numerous congressional debates and the key 1975 *National Resources Defense Council* v. *Calloway* decision, "navigable waters" were expanded to what is currently referred to as "Waters of the United States" (Lewis, 2001). The USACE defined "waters of the United States" to mean "all waters which are currently used, or were used in the past, or may be susceptible to use in interstate or foreign commerce, including all waters which are subject to the ebb and flow of the tide" (Dennison and Berry, 1993). In addition, the definition also includes "all other waters such as intrastate lakes, rivers, streams (including intermittent streams), mudflats, sandflats, wetlands, sloughs, prairie potholes, wet meadows, playa lakes, or natural ponds, the use, degradation or destruction of which could affect interstate or foreign commerce" (33 CFR 328.3).

Several other court cases throughout the late 1970s and early 1980s upheld this definition, including: *U.S.* v. *Ashland Oil & Transportation Co.* in 1974, *United States* v. *Byrd* in 1979, *United States* v. *Riverside Bayview Homes Inc.* in 1985, and *Hoffman Homes* v. *Administrator* in 1993 (Downing *et al.*, 2003: 475–493). However, the broad definition of wetlands under federal jurisdiction was not permanent. Perhaps the most recent and sweeping ruling on the federal jurisdiction of wetlands was decided by the 2001 U.S. Supreme Court ruling of the *Solid Waste Management Agency of Northern Cook County* v. *U.S. Army Corps of Engineers.* The decision focused on isolated, non-navigable intrastate wetlands which had been previously protected by Section 404. The majority of the court ruled that Congress had not clearly expressed its intent to regulate such waters. The "SWANCC" decision was considered to be a departure from prior wetland jurisdiction decisions and has provided opportunities to question many broader issues of CWA jurisdiction (Downing *et al.*, 2003: 475–493). Although the decision only affected isolated, non-navigable intrastate wetlands and upheld the protection of wetlands adjacent to navigable waters, numerous lawsuits since the decision have challenged the jurisdiction over other types of waters that are nonisolated.

The definitions of key terminology and judicial interpretations of federal jurisdiction are the heart of Section 404 and its application to wetlands. Without these interpretations, wetlands protection and the CWA as a whole would be very limited in geographic scope. As outlined above, the interpretations of the definition of "navigable waters" and "waters of the United States" are the critical link to federal wetland protection. When viewed as a whole, federal protection of wetlands has no doubt increased since 1972, but the judicial interpretations that grant this protection are subject to change at any time.

Implementing Section 404: The U.S. Army Corps of Engineers

As described above, the USACE was the agency charged with the responsibility of issuing permits, which were for the "discharge of dredged or fill material" into "waters of the United States," with the EPA retaining oversight and veto power over permit decisions. With some trepidation, the USACE began its permitting process in 1975. Because its jurisdiction was so far-reaching under regulatory interpretations, the permitting program was implemented in three phases. July of 1975 saw the implementation of the first phase of the permitting program, with its jurisdiction applying to coastal waters, navigable inland rivers and lakes, and wetlands adjacent to these waters. The second phase began in September of 1976 and added all lakes, primary tributaries, and their adjacent wetlands. Finally, in July of 1977 the USACE added all remaining jurisdictional waters including isolated wetlands.

However, phase 2 was implemented too early, as USACE did not have enough "regulatory resources" to cover the entire scope of CWA jurisdiction, such as intrastate water bodies and smaller streams above the headwaters of rivers (Downing *et al.*, 2003: 475–493). Due to the lack of regulatory resources, the USACE implemented a system of "General" permits to be issued for activities thought to have very limited potential for detrimental environmental impacts. This attempt to streamline the permitting process has evolved into two categories of USACE Section 404 permits: standard and general. The conditions under which each type of permit is issued varies by the type of activity, the impact of the activity, and the district or region where the activity will be located. Currently, the USACE issues four types of Section 404 permit within the two categories: standard permits include

Individual permits and Letters of Permission; general permits include Regional General permits and Nationwide permits.

Individual permits

Individual permits are the basic and original form of authorization used by USACE districts. Activities that entail more than minimal impacts require an individual permit. Processing the permits involves evaluation of individual, project-specific applications in what can be considered three steps: pre-application consultation (for major projects), formal project review, and decision making.

Once a permit application is submitted, the USACE must inform the applicant of any deficiencies in the application within 15 days. Once the applicant has supplied all required information, USACE determines if the application is complete. Within 15 days of that determination, USACE must issue a public notice of the application for posting at governmental offices, facilities near the proposed project site, and other appropriate sites. In the public notice, USACE requires that any comments must be provided within a specified period of time, typically 30 days (33 CFR 325.5b). When determining if the activity is necessary, the engineers at USACE must consider whether the activity is dependent on being located in the wetland, or if alternative sites are feasible. If the applicant can show that no practical alternatives exist, then the activities must be performed to minimize adverse impacts to the wetland. The applicant must also provide compensation for any unavoidable impacts, typically carried out through some form of mitigation. The USACE evaluates the public benefits and detriments of each case.

Relevant factors considered by USACE include: conservation, economics, general environmental concerns, aesthetics, wetlands, floodplain values, cultural values, navigation, fish and wildlife values, water supply, water quality, and any other factors judged important to the needs and welfare of the people (Connally *et al.*, 2005). In addition, individual state permitting and water quality certification requirements can provide an additional safeguard to the USACE permitting program. Section 401 of the Clean Water Act requires state certification or waiver of certification prior to issuing an Individual Section 404 permit.

Letters of Permission

The first alternative form of authorization used by the USACE for certain prescribed situations is the Letter of Permission. Letters of Permission may be used where, in the opinion of the district engineer, the proposed work would be minor, not have significant individual or cumulative impact on environmental values, and should encounter no appreciable opposition (33 CFR 325.5b2). In such situations, the permit application is coordinated with other relevant agencies, as well as adjacent property owners who might be affected by the activity. Public notices and comment periods are not required for this permit type.

A Letter of Permission can be issued much more quickly than a standard Individual permit, since many of the Individual permit requirements are bypassed. Any project the USACE proposes to authorize under a Letter of Permission may be elevated to an Individual permit by the EPA or state department of environmental management (or equivalent).

Regional General permits

As noted above, general permits arose from a lack of regulatory resources during the final phase of Section 404 implementation and represented an attempt to streamline the permitting process for common activities. The USACE considers both Regional General permits and Nationwide permits to be "general" permits. But, because they can differ across USACE districts, we address them as separate types. General permits are issued when "activities are substantially similar in nature and cause only minimal individual and cumulative impacts" (USACE, 2001). This permit type covers activities in a limited geographic area, or a region of the country. General permits are reviewed every five years and an "assessment of the cumulative impacts of work authorized under the general permit is performed at that time if it is in the public interest to do so" (USACE, 2001). In developing general permits, the USACE must go through a public interest review and receive certification by the state in which the permit is being processed. Once a general permit is issued for an activity, individual projects meeting the terms and conditions of the general permit category can quickly receive authorization without additional review by the USACE.

Nationwide permits

Nationwide permits are a special type of general permit. Activities covered under Nationwide permits can go forward without further Corps approval as long as the conditions set forth in the Nationwide permit category of work are met. By far the most commonly issued type, Nationwide permits are issued for specific activities that are deemed to have "no more than minimal adverse effects on the aquatic environment, both individually and cumulatively" (Issuance of Nationwide Permits Notice, 2005: 2019–2095). The activities allowed under Nationwide permits are broad and have been the source of criticism by environmental groups in the past for serving as a loophole for Section 404 permitting. Unlike Regional General permits, the work allowed by Nationwide permits applies, as the name suggests, to the entire nation.

Difficulties of wetland mitigation

The four permit types discussed above are the key means by which the USACE manages discharges into waters of the U.S. and, due to federal judicial interpretations, wetland alterations and losses. Previous research pertaining to Section 404 is limited in scope. Many studies on Section 404 permitting are focused primarily on wetland mitigation: creating, restoring, or protecting wetland losses as a result of Section 404. Unfortunately, the results of this body of research do not provide support for successful wetland mitigation programs. For example, in Louisiana, 41% of permits issued between 1982 and 1986 required mitigation, but only 8% of the total area was mitigated (Sifneos *et al.*, 1992). Other studies suggest that between 17% and 34% of restored or created wetland projects had not been constructed at all (Kusler and Kentula, 1990; Owen and Jacobs, 1992: 345–353).

For example, research that included Texas found a net loss of 917 acres (371 hectares) of wetlands in the USACE Fort Worth District between 1982 and 1986 for permits that did not fulfill their required compensatory mitigation (Sifneos *et al.*, 1992). Comparable results were found in Oregon and Washington. Kentula *et al.* (1992) found that over a 10-year period in Oregon (1977–1987), 183 acres of wetlands were impacted and 111 acres (45 hectares) were created – a 43% net loss. In Washington from 1980 to 1986, 151 acres (61 hectares) of wetlands were impacted and 112 acres (45 hectares) were created, amounting to a 26% net loss. Permitted activities in both states occurred near urban areas experiencing outwardly sprawling development patterns (Kentula *et al.*, 1992: 109–119). Owen and Jacobs (1992) conducted a similar study in Wisconsin, finding in the first six months of 1988, 422 acres (170.7 hectares) of wetlands were allowed to be filled, while only 40 acres (16 hectares) were actually created.

Compensatory mitigation required under some Section 404 permits frequently involves creating new wetlands. However, these created wetlands do not often achieve the same degree of ecological functionality as natural wetlands, even several decades after they are created (Campbell *et al.*, 2002: 41–49; Cole and Brooks, 2000: 221–231; Cole and Shafer, 2002: 508–515). This is due to their creation in inappropriate hydrologic conditions or an inadequate program to monitor the progress of the mitigated wetland ecosystems over time (Cole and Brooks, 2000: 221–231; Cole and Shafer, 2002: 508–515; Gallihugh and Rogner, 1998). Further, constructed wetlands are typically not capable of replacing the functionality of the lost wetland because mitigation does not always require restoration or creation of the same wetland type being converted (Cole and Shafer, 2002: 508–515). Finally, mitigation projects may be established far from the location of original impact. Consequently, functions added by constructed wetlands have been transferred from an environment where their functions were important to an environment where they no longer have the same value. This spatial pattern is particularly evident in Florida, where mitigation banks have been created inland to compensate for wetlands altered in coastal areas where property is far more valuable.

Other research demonstrates a chronic lack of oversight concerning compensatory mitigation within the USACE. For example, a comprehensive report on wetland mitigation written and assembled by the NRC (2001) found that inspections to ascertain compliance with Section 404 mitigation requirements were rarely conducted. Reviewing applications and granting permits often took precedence over the vast majority of required mitigation reviews (NRC, 2001). Most recently, the Government Accountability Office (GAO) reported that only 21 of the 89 permit files randomly selected in its study contained the required evidence of monitoring restoration activities (GAO, 2005). Further, only 15% of the permit files contained evidence that the USACE had conducted required compliance inspections.

These overall findings are similar in both the Galveston and Jacksonville Regulatory Districts of the USACE, which encompass coastal Texas and the entire state of Florida. In the Galveston District, of the 18 Individual permits that were reviewed, 11 required monitoring reports. However, only one permit record actually contained the report. In the Jacksonville District, 11 of the 16 Individual permits contained the required monitoring reports (GAO, 2005). Although lack of oversight does not directly indicate that mitigation is not occurring (successfully or otherwise), it does suggest that compensatory mitigation is not achieving the desired goal of reducing overall wetland loss.

The impact of Section 404

The Status and Trends reports, mandated by the 1986 Emergency Wetlands Resources Act, offer the most thorough examination of wetland loss in the U.S. The 1990 report, which examined the 1970s to 1980s, showed large declines in wetlands in the conterminous U.S.: 2.6 million acres (1 052 182 hectares) of wetlands were lost during the study period, an annual average loss of 290 000 acres (117 358 hectares) (Dahl and Johnson, 1991). Future reports showed even greater declines. From 1986 to 1997, an estimated 644 000 acres (260 617 hectares) of naturally occurring wetlands were converted, an annual average loss of 58 500 acres (23 674 hectares) (Dahl, 2000). In contrast, the most recent Status and Trends report covering the years 1998 to 2004 found a reversal of the pattern of wetland loss across the U.S. The study reports that there was a net *gain* of wetlands during the study period, with the U.S. adding an additional 191 750 acres (77 598 hectares) of wetlands, an average gain of 32 000 acres (12 949 hectares) per year (Dahl, 2006: 112). This initial result is, however, quite misleading. During the study period, the report classified over 700 000 acres (283 279 hectares) of ponds and other water habitats as wetlands. In reality, over 500 000 acres (202 342 hectares) of truly naturally occurring wetlands were lost (Dahl, 2006: 112).

Another smaller body of research on Section 404 permitting is centered on pre-post permit landscape conditions and associated cumulative impacts. For example, Stein and Ambrose (1998) conducted an on-site study examining riparian areas in the Santa Margarita watershed in Southern California. They concluded that while the Section 404 program had reduced overall project impacts, it had not minimized cumulative impacts. The authors also examined Nationwide permits and found they accounted for proportionally more cumulative impacts, despite the fact that they affect less total area. Research on Section 404 impacts in North Carolina revealed that not only were naturally occurring wetlands lost under the USACE permitting program, but habitat fragmentation occurred in 80% of areas adjacent to permit sites (Kelly, 2001: 3–16). This finding suggests the presence of additional "nibbling" impacts associated with permitted activities that are not taken into consideration during initial review.

Section 404 in Texas and Florida

Our research builds on previous studies by examining the spatial pattern of Section 404 permits in coastal Texas and Florida. Specifically, we analyze the record of USACE permits over a 13-year period to better understand the extent of wetland alteration, the types of wetlands being permitted, and the effects of wetland loss on flooding (based on streamflow rates) and flood damage (based on property loss estimates) within our study areas.

Texas and Florida are unique cases in terms of wetlands and wetland loss. Both states rank among the top five in the U.S. in terms of wetland area and human population. However, Florida has experienced much higher levels of development in coastal areas that are vulnerable to wetland loss. The Texas coast has not been developed to the same extent as Florida, yet projections indicate that coastal urban and suburban development will occur in the near future. Examining Section 404 permit activity in both states through the use of descriptive statistics and spatial analysis provides an initial foundation for exploring the scope of wetland alteration and the importance of wetlands for mitigating the adverse impacts of floods.

Spatial pattern of Section 404 permits in Texas and Florida

Over the 13-year period from 1991 to 2003, a total of nearly 46 000 Section 404 permits were issued in the Galveston District in Texas and the Jacksonville District in Florida. After cleaning both datasets provided by each USACE district of data entry errors, duplicate entries, permit renewals and changes in permit type, we

Table 6.1 *Section 404 permits issued in Florida and Texas from 1991 to 2003 by permit type*

| | Permit type | | | | | | | |
| | General | | Individual | | Letter of Permission | | Nationwide | |
State	Total	Percent of total	Total	Percent of total	Total	Percent of total	Total	Percent of total
Texas	3 512	31.50	1 284	11.50	1 237	11.10	5 116	45.90
Florida	4 963	18.10	3 959	14.40	2 027	7.40	16 505	60.10
Total	8 475	22.00	5 243	13.60	3 264	8.50	21 621	56.00

Adapted from Brody *et al.* (2008) "A spatial-temporal analysis of section 404 wetland permitting In Texas and Florida: thirteen years of impact along the coast," *Wetlands*, **28**, 107–116, with permission.

derived for analysis a dataset of 38 603 Section 404 permits, 71% of which were issued in Florida. This is not a surprising result given the rapid growth and development taking place in Florida during the study time period.

A breakdown of permits by type during the study period shows similar patterns within each state. Individual permits and Letters of Permission are the least-issued permit types (Table 6.1). In contrast, Regional General and Nationwide permits are by far the most issued permit types in both areas when viewed together. These two types make up 64.5% of the permits issued in Texas and Florida, which makes sense, considering the lack of administrative hurdles in obtaining either type. The primary difference between the study states is the percentage of Nationwide and General permits. Texas appears to have a greater reliance on the General Regional permits (31.5%) relative to Florida (18.1%), while Florida has a greater reliance on the Nationwide permit mechanism (60.1%) relative to Texas (45.9%). This difference is likely due to two factors. First, the bulk of Texas' General permit categories is related to oil and gas activities in specific areas, making the General permit more attractive for extraction-based industry. Second, Florida has a larger percentage of wetlands compared with other areas in the U.S., tilting the activities allowed under Nationwide permits in the state's favor.

Although examining permit counts by type can be instructive in gauging the extent of wetland alteration across coastal Texas and Florida, linking permits to specific spatial characteristics can be even more revealing about the location of the impacts. Using GIS and associated spatial analytical procedures, we tied our Section 404 permit database to locations relative to urban areas and 100-year floodplains (for more information see: Brody *et al.*, 2008a: 107–116). The results of this analysis

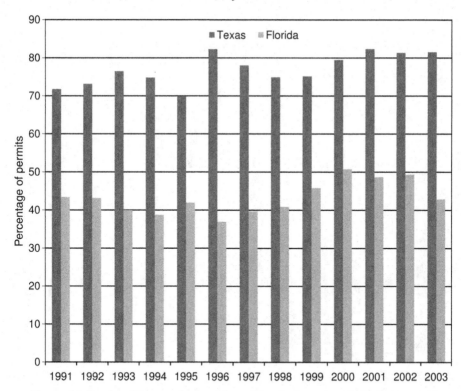

Figure 6.1 Section 404 permits issued outside of urban areas in Galveston and Jacksonville district USACE study areas, 1991–2003.

show markedly different patterns across the two study areas. For example, in coastal Texas, an average of 78% of permits was issued outside of urban areas during the study period. In Florida, this figure dropped to just over 57%. When viewed on an annual basis, permit counts outside urban areas in Texas generally increased, while in Florida the same annual pattern was generally stable (Figure 6.1).

The difference in these spatial patterns is likely due to the planning climates, or tolerances, between the two states. Coastal development in Texas is more likely to take place in once rural settings (of which there are plenty) in a "leapfrog" fashion due to a lack of land use planning and development management policies. Conversely, development in Florida takes place under a statewide planning mandate involving specific land use controls, which most likely reduces the ability to develop wetlands outside city limits. Southern Florida also has established large protected areas in the southern part of the state, which helps constrain development within preexisting urban cores (see Brody *et al.*, 2006b: 75–96).

The pattern of Section 404 permits issued within 100-year floodplains in the study areas also show differences. In Florida an average of 48% of wetland alteration permits were issued within floodplains, compared to an average of 39% in the

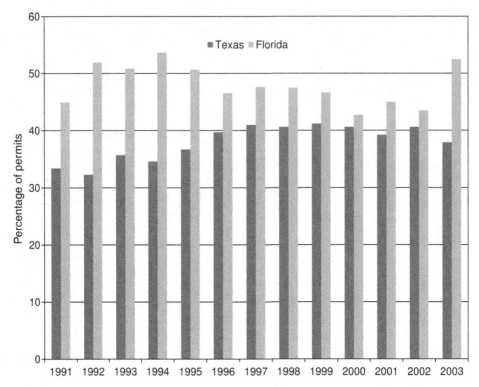

Figure 6.2 Section 404 permits issued within 100-year floodplains in Galveston
and Jacksonville district USACE study areas, 1991–2003.

Texas study area. When analyzed annually, the differences both between and within
the two study areas is fairly constant (Figure 6.2). From 1991 to 2003, permits issued
within floodplains in Florida ranged from a low of 42.7% to a high of 53.7%. The
results for the Texas study were similar, ranging from a low of 32.2% to 41.2%.

The fact that Florida had, year after year, more Section 404 permits issued in the
floodplains is not surprising; the state had more area in designated floodplains and saw
more widespread development over the period of record we examined. It should also
come as no surprise that permits to alter or fill wetlands and other areas that are determined to be "waters of the United States" occur in floodplains. After all, floodplains
are often characterized as wet areas, at least intermittently. It is, however, a great concern that development and other projects are consistently allowed to decrease the flood
mitigation effects of wetlands in areas where they are most needed.

Section 404, flood damage, and streamflow

It is often remarked that wetlands reduce floods, and the empirical literature in
Chapter 5 certainly supports this statement. It should follow then, that a loss of

wetlands or their function increases flood events, both in frequency and magnitude. However, relative to studies investigating the role of wetlands and water quantity, there is a dearth of empirical research examining wetland *loss* and flooding. This is at least in part due to data constraints, as it is difficult to determine when a wetland was lost and then tie this loss to water quantity metrics. However, the use of Section 404 permits as a measure of wetland alteration overcomes this problem.

To date, we have conducted five research studies investigating the effects of Section 404 permits on water quantity using measures of streamflow and estimated flood damage (Table 6.2). These studies, with little variation, have found Section 404 permits to have positive and statistically significant effects on flooding and flood impacts. The results hold despite varying units of analysis, changing study periods, and different forms of measuring both flooding and Section 404 permits. Most importantly, these results continue to hold after statistically controlling for additional climatic, hydrologic, socioeconomic, and policy-related variables (see Table 6.2).

As mentioned above, streamflow measurements collected from USGS gauging stations enabled us to examine the effects of Section 404 permits. The first of several studies spanned coastal Texas and all of Florida, utilizing hydrologic units established by the USGS (Brody *et al.*, 2007b: 413–428). A total of 85 hydrologic units were incorporated into the analysis, 39 in Texas and 46 in Florida. Average monthly streamflows were calculated for every gauge in each hydrologic unit over a 12 year study period from 1991 to 2002. Counts of "exceedances," or the number of months that a gauge surpassed the study period average, were calculated and averaged within each unit. Results showed that issued Individual and Regional General Section 404 permits were positive and statistically significant drivers of flow exceedances. These relationships held even after controlling for a host of additional variables, including hydrologic unit area, slope, total length of streams in each unit, impervious surface area, the number of dams, population density, and median household income (Brody *et al.*, 2007b: 413–428)

We conducted a second study using streamflow measurements that focused solely on Texas. In this analysis, all USGS stream gauge locations within the 49 county USACE Galveston District were selected for the period 1996–2003 (Highfield, 2008) peak annual flow, log-transformed to approximate a normal distribution. Instead of using established hydrologic units, sub-basins were delineated around each of the 47 stream gauge locations. We accounted for temporal considerations by using a cross-sectional time-series (CSTS) approach to analysis.

Across four CSTS models, separate models for each permit type – Individual, Nationwide, Letter of Permission, and Regional General – were all positive and significant with respect to increasing peak annual flow. Once again, issued wetland alteration permits proved significant factors in increased peak flow even after controlling

Table 6.2 *Summary of studies investigating the relationship between Section 404 permits and flood metrics.*

Dependent variable	Section 404 permit measurement	Direction (significance)	Study area (study period)	Citation
Counts of exceedances over study period average	Permits by type	Individual (positive[*]) General (positive[**]) Nationwide (positive) Letter (negative[**])	Coastal watersheds in Texas and Florida (85)	Brody *et al.*, 2006a:294–310
Flood damage (dollar amount)	Permits by type in/out of floodplains	Individual (positive[**]) Nationwide (positive) All permits out (negative[**])	Florida (1997–2002)	Highfield and Brody, 2006: 23–30
Flood damage (dollar amount)	Total cumulative permits	Positive[**] (p<0.01)	Florida (1997–2001)	Brody *et al.*, 2007a: 330–345
High damage event, dichotomous	Total cumulative permits	Positive[*] (p<0.05)	Florida (1997–2001)	Brody *et al.*, 2007a: 330–345
Flood damage (dollar amount)	Total cumulative permits	Positive[**] (p<0.01)	Coastal Texas (1997–2007)	Brody *et al.*, 2007b
Maximum daily streamflow	Total cumulative permits by type	Individual (positive[**]) General (positive[**]) Nationwide (positive[**]) Letter (negative[**])	Coastal Texas sub-basins (1996–2003)	Highfield, 2008

*p<.05 **p<.01

for variables, including: sub-basin area, shape, and slope, natural and impervious surface land cover, and soil permeability. Not only did permits have a significant positive effect on peak annual flow, but we found larger areas of wetlands in each sub-basin actually worked to reduce peak annual flow. More specifically, increasing areas of Palustrine scrub/shrub wetlands significantly reduced flooding. These results not only confirm the role of wetlands in reducing peak streamflow found in previous research, they simultaneously demonstrate the effects of altering or removing wetlands on increasing peak flows and potential damages from flooding.

While streamflow is an objective measurement of water quantity and potential flooding, it does not address visible impacts that often drive local policy decisions. Thus, we conducted a second wave of analysis that examines the impacts of wetland alteration on the results of flooding: property damage. First, we examined issued Section 404 permits in FEMA-delineated 100-year floodplains at the county level in Florida (Highfield and Brody, 2006: 23–30). Results showed that alteration of wetlands inside the floodplain led to significantly higher amounts of flood damage when controlling for precipitation, median structure improvement value, and population density. Perhaps the most important result of this research in terms of Section 404 activity and flooding was the role of Individual permits. When looking at standardized regression coefficients, Individual permits (0.48) had the highest impact on explaining flood damage – even higher than the amount of precipitation (0.41). This finding demonstrates the potential for naturally occurring wetlands to mitigate the adverse impacts of floods.

We conducted a related analysis using the gross number of Section 404 permits measured cumulatively from 1997 to 2001 in both coastal Texas and Florida. In Florida, flood damage at the county level (the smallest available spatial unit for measuring property damage at the time) was predicted for individual events by cumulative counts of wetland permits, along with a diversity of geophysical-, socioeconomic-, and planning-related variables (Brody *et al.*, 2007a: 330–345). Despite numerous statistical controls, Section 404 permits remained positive and significant predictors of reported flood damage. In fact, wetland alteration permits had a stronger statistical effect (in terms of standardized regression coefficients) than impervious surface area, dams, floodplain area, flood duration, stream density, and housing value density. Not only did increasing numbers of permits lead to more damaging floods, they also had a positive effect on "high damage" flood events. That is, Section 404 permits had positive and significant impacts on floods that exceeded the aggregate median value of $50 000, even after controlling for the same groups of variables described above (Brody *et al.*, 2007a: 330–345).

In the aggregate, the price of a permit in terms of corresponding flood damages to homes and business is quite high. In fact, based on our model, each issued permit in Florida in the period 1997–2001 resulted in almost $1000 in added property damage per flood. When considering the number of permits issued, that equals over $402 465 per flood, or about $30 426 354 per year in added damage for the entire state.

Strikingly similar relationships held when conducting this analysis in coastal Texas. Nearly the same set of statistical control variables were used to analyze SHELDUS-derived flood damage data for 37 counties in the coastal region (Brody *et al.*, 2008b: 1–18). In this analysis, only the amount of precipitation the day before a flood event and the duration of the flood itself were statistically stronger predictors than Section 404 permits. Wetland loss, as measured through permits, was

statistically a stronger factor in predicting flood damage than the amount of precipitation the day of the flood event, floodplain area, area of impervious surface,
number of dams, CRS rating, and median household income within each county
in the sample.

What is the overall price of wetland permits in coastal Texas? According to
our model, one permit to alter a naturally occurring wetland equals an average of
$211.88 in added property damage per flood. This figure does not sound like that
much, but consider the fact that thousands of permits are issued in this area each
year and that number is increasing over time. Just a five-year glance at the flood
problem in coastal Texas (1997–2001) shows that wetland permits equated to over
$38,000 in added property damage per flood.

Summary

Across all of the research regarding Section 404, several commonalities and
important implications come to light. First, aggregate wetland permits are consistently positive and significant in explaining both streamflow and flood damage
estimates – in both coastal Texas and Florida. This finding confirms that not only
do wetlands reduce floods, but Section 404 permits serve as indicators of wetland
loss. Second, when permits are treated and analyzed separately, Individual permits
always have the greatest effect on flooding. This is an expected result since they are
likely to represent the largest wetland losses, but important because it confirms the
role of Section 404 permits as an indicator of wetland loss and as having a positive
effect on flooding and associated flood damages.

The evidence above clearly illustrates the need to incorporate wetland protection and preservation as a key tool in effective flood mitigation. Whatever specific
statistical model we choose to analyze, the result is always the same: that the alteration of naturally occurring wetlands in Texas and Florida significantly increases
flooding events and associated property damage. While policies have long been in
place to protect wetland values for wildlife and water quality, federal wetland regulations specific to flood control have been absent. However, some attempts have
been made. Industry groups such as ASFPM and FEMA's CRS have highlighted
and encouraged wetland protection for flood mitigation. However, these programs
are voluntary and are likely not sufficient to create the lasting, informed, and geographically broad policies necessary to achieve the goal of wetland preservation.

7

Mitigation strategies and reduction of flood damages

Non-structural forms of mitigation, as described in Chapter 5, may be the most overlooked, yet promising approach to reducing the adverse effects of chronic flooding at the local level. Mitigation techniques via land use planning, education, training, etc. have been advocated by researchers for decades, yet remain virtually untested in the planning and management literature. This chapter addresses the issue by examining the impact of various non-structural mitigation strategies on the severity of losses caused by floods in Texas and Florida. First, we investigate the effectiveness of FEMA's CRS in reducing property damage and human casualties within the study's states. Second, we analyze specific mitigation techniques in more depth using a survey of floodplain administrators and local planners. Results show the extent to which non-structural measures are adopted throughout the study area and the degree to which they reduce adverse impacts from floods. These findings provide guidance to local decision makers in coastal regions on how to establish programs that foster flood-resilient communities.

FEMA's CRS as a vehicle for local flood mitigation

As detailed in Chapter 4, FEMA's CRS is meant to encourage local jurisdictions to exceed the NFIP's minimum standards for floodplain management. Participating communities implement flood mitigation measures in exchange for national flood insurance premium discounts of up to 45%. Credit points are assigned for 18 different flood mitigation activities falling within four designated series (see Table 4.1 in Chapter 4). Credit points are aggregated into classes, from lowest (9) to highest (1). Communities awarded a higher CRS class will have implemented a greater number of the 18 flood mitigation measures and therefore receive a higher premium discount for insurance coverage. While many consider the CRS a perverse incentive because it makes it less expensive to develop and live in the 100-year floodplain, it nevertheless better prepares communities for the adverse impacts of floods. The

CRS is also an ideal indicator of the degree to which a locality is mitigating against flooding through primarily non-structural techniques, because a policy must be implemented to receive credit. Every participating community's program is monitored, reviewed, and updated on a yearly basis to ensure compliance and enable jurisdictions to improve their efforts and increase their scores (see Chapter 4 for more details).

Does the CRS work?

While FEMA's CRS has become an established program based on known non-structural mitigation techniques, there has never been a systematic, program-wide analysis of its effectiveness beyond a single community since its inception in 1990. How can FEMA justify the continuation of this program during a time of scarce financial resources and continued increases in flood losses? What information helps non-CRS communities make a decision to participate in the program? In light of these unanswered questions, we examined all 67 counties in Florida and 37 coastal counties in Texas over a five-year period to observe whether the CRS program is achieving its intended goals. In Florida, from 1997 to 2001, we catalogued 383 flood events causing over $979 million in reported property damage, for an average of $2 638 712 damage per flood (Brody *et al.*, 2007a: 330–345). In a companion study in Texas, we recorded 423 flood events responsible for over $320 million in reported property damage among counties in the coastal region (Brody *et al.*, 2008b: 1–18); the average damage per flood was $423 766.

By correlating the class of participating CRS jurisdictions with reported property damage on a per-flood basis, while controlling for multiple natural environment, built environment, and socioeconomic contextual characteristics, we discovered that CRS participation has resulted in significant flood damage reduction in both states. In Florida, results based on multivariate regression models (Table 7.1) show a real unit change in CRS class (moving in increments of 5%) equal to a $303 525 decrease in the average amount of damage per flood. Based on the parameters of our statistical model for Florida, the CRS rating is more than twice as effective as dams in reducing flood property damage. Putting these results into a climatological context, the property damage saved by a 5% increase in the CRS discount for insurance premiums is roughly equal to the added amount of property damage associated with 2 inches (5 cm) of precipitation (see Brody *et al.*, 2007a: 330–345 for more details). In fact, based on the standardized regression coefficient in Table 7.1, the statistical effect of the CRS on reducing flood damage is the most powerful predictor among all variables in the model, except for the amount of precipitation and damage in adjacent counties (included to statistically control for neighboring effects).

Table 7.1 *OLS regression models predicting property damage from floods in Florida, 1997–2001*

	b	Beta
Socioeconomic baseline variables		
Adjacent damage	0.12 800**	0.31 007
	(0.02 155)	
Housing value density	0.02 517*	0.10 654
	(0.01 216)	
Biophysical variables		
Precipitation	0.06 353**	0.23 260
	(0.01 628)	
Floodplain area	30.65e–10*	0.13 744
	(1.60e–10)	
Flood duration	0.02 336*	0.14 728
	(0.01 105)	
Stream density	0.12 890	0.05 122
	(0.14 516)	
Planning decision variables		
Impervious surface	–8.52e–11	–0.01 792
	(3.64e–10)	
Wetland alteration	0.00 038**	0.15 071
	(0.00 011)	
Dams	–0.00 273	–0.07 122
	(0.00 172)	
FEMA CRS rating	–0.02 331**	–0.15 105
	(0.00 910)	
Constant	3.74 624**	
	(0.19 632)	
N	367	
Probability > F	0.000	
R2	0.2812	
√MSE	0.99 208	

Robust standard errors are in parentheses. Null test of coefficient equal to zero.
*p<0.05, **p<0.01.
MSE, mean squared error; OLS, ordinary least squares.
Modified from Brody *et al.* (2007a) "The rising costs of floods: examining the impact of planning and development decisions on property damage in Florida," *Journal of the American Planning Association,* **73** (3), 330–345, with permission.

Empirical results from our Texas study reveal a similar pattern of flood damage reduction. Among coastal counties in Texas, from 1997 to 2001, a real unit increase in CRS classes equaled a $38 989 reduction in the average property damage per flood. Theoretically, if every jurisdiction in the study area had maximized their

CRS rating (e.g., achieved a class of 1), the cost of floods would have been less than a quarter of the $320 million catalogued (Brody *et al.*, 2008b: 1–18). While this calculation may seem like semantics, since it is highly unlikely coastal counties in Texas will ever achieve the highest possible CRS class, it demonstrates the effectiveness of non-structural flood mitigation techniques in terms of reducing the severity of flood damage over time. Furthermore, the CRS appears to not only save property, but also human lives during flood events. For example, a sister study covering 99 counties in the eastern portion of Texas from 1997 to 2001 found that for every real unit increase in CRS class, the odds of death and injury from a flood event decrease by over 36% (Zahran *et al.*, 2008: 537–560).

The results of our statistical inquiry clearly indicate the usefulness of the CRS in reducing the adverse impacts of floods. Our findings give credence to the notion of providing the public with sound information about floods, implementing land use planning policies such as open space protection and land acquisition, and, finally, being well prepared with warning and safety protocols. So, why are there only about 1100 communities participating in the CRS nationwide? Although there is no single factor explaining the lack of participation in the CRS, participation rates most likely stem from a complex set of interrelated issues that include the following. First, the perceived threat of flooding in most local jurisdictions is low compared with more immediate issues facing local governments, such as economic development, schools, and crime. Given scarce resources and personnel, localities may opt not to make the effort to join the program. Second, localities often avoid federal oversight or control unless a program is mandatory. The CRS requires monitoring and approval at the federal level in regulatory areas usually controlled by local governments. Third, the CRS involves a somewhat arduous accounting and monitoring system that must accommodate regular assessments. As already indicated, from the perspective of a local jurisdiction, such an effort may not be worth the potential benefits. Fourth, actually implementing various activities is expensive, time-consuming, and possibly politically contentious even though it could result in considerable cost-savings for residents over the long term.

Finally, because the CRS program acts as somewhat of a perverse incentive to develop and reside in areas vulnerable to flooding, communities may choose not to participate. Insurance premium discounts earned through community-wide mitigation activities make it less expensive, and therefore more encouraging, for homeowners to live in the 100-year floodplain (where there is a 1% chance of flooding every year). Thus, the CRS may actually facilitate development in areas most vulnerable to flooding even though residents living in these areas are better prepared to address flood events. Communities may choose not to participate in the CRS because the benefits of flood mitigation activities may be outweighed by the risk of more people living in the floodplain. A more viable strategy for facilitating resilient

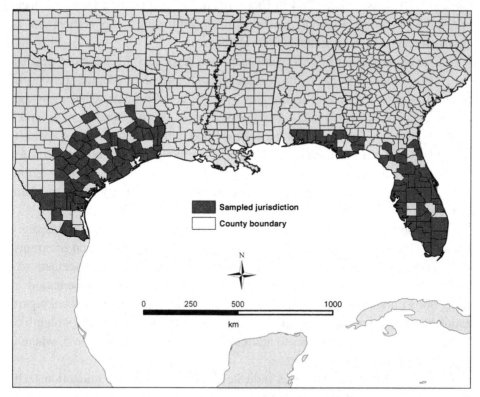

Figure 7.1 Survey localities in Florida and Texas.

communities over the long term may be to adopt land use planning techniques that explicitly direct growth away from areas vulnerable to chronic flooding.

Beyond the CRS: examining specific local mitigation techniques

While the CRS is a strong indicator of local flood mitigation, it provides only a starting point for understanding the extent to which mitigation techniques are embedded into local planning and decision-making frameworks. To understand in more detail the state of flood mitigation and its potential effectiveness, in 2005 we conducted a survey of localities with 5000 or more residents throughout our study area. Our sampling strategy covered all of Florida, and in Texas local jurisdictions intersecting fourth-order hydrological units (as defined by USGS) and located within 100 miles (161 km) of the coastline. As seen in Figure 7.1, this strategy enabled us to evaluate a geographically representative sample of jurisdictions.

A self-administered web-based questionnaire was distributed to each jurisdiction in the sample. We targeted the planning directors (or lead planners) in Florida, and floodplain administrators, the administrative equivalent in Texas. Overall, we contacted 471 jurisdictions: 264 in Florida and 207 in Texas. In total, 173 local jurisdictions responded to our questionnaire, resulting in a cooperation rate of approximately 35% in Florida and 38% in Texas.

The survey consisted of two main parts (see Appendix 7.1 for more details). First, we questioned localities on the degree to which they adopted a series of structural and non-structural techniques over the last five years. Questions were answered on a scale from 0 to 2, where 0 is never used, 1 is used occasionally, and 2 is used extensively. Of the jurisdictions using flood mitigation techniques, approximately 60% did not participate in the CRS, even though most would qualify for the program and some level of insurance premium discount. This result is telling of how resistant localities can be to federal oversight, particularly in Texas where local control over development is a paramount issue.

The second part of the survey focused on aspects of organizational capacity thought to increase the likelihood of the adoption of strong mitigation programs, such as commitment, staffing, financial resources, and adaptive management. As discussed in Chapter 5, previous research suggests that the implementation of strong flood mitigation policies is largely driven by the capacity of the local organization administering the program. Respondents were asked to rate the strength of 15 organizational characteristics to address floods on a scale from 0 to 5, where 0 is not present and 5 is very strong.

Analysis of frequency statistics indicate the extent to which mitigation techniques have been adopted across Florida and coastal Texas. As shown in Table 7.2, clearing debris that may block channels or drainage devices is the most extensively used local strategy, with over 65% of surveyed localities using this mitigation strategy extensively. Retention or detention ponds associated with new developments are also widely used techniques among both Texas and Florida localities. Fifty% of the sample reported they use this technique extensively. More expensive politically contentious measures, such as channelization, dams, and levees are less commonly adopted. For example, 50% of localities report never using channelization of watercourses to structurally mitigate flood risks (Brody *et al.*, 2009a: 492–515).

Of the 14 non-structural flood mitigation strategies examined in the survey, land development codes are used most extensively, reported by over 67% of surveyed localities (87.3% in Florida and 43.8% in Texas). These codes serve as the regulatory foundation for development management by setting specific standards for zoning, site development, subdivisions, environmental issues, etc. For example, the St. Johns County's, Florida, land development code contains provisions to prohibit certain uses within the 100-year floodplain. Similarly, Manatee County, Florida, has adopted subdivision regulations pertaining to flood hazards. Construction codes designed to address flooding and severe storms is another frequently employed technique, used extensively by almost 64% of responding localities.

Table 7.2 *Frequency and descriptive statistics of flood mitigation techniques*[a]

	Never used	Moderately used	Extensively used
Structural strategies			
Retention/detention	28	23	51
	(27.5)	(22.5)	(50.0)
Levees	83	10	5
	(84.7)	(10.2)	(5.1)
Channelization	50	22	28
	(50.0)	(22.0)	(28.0)
Dams	79	14	3
	(82.3)	(14.6)	(3.1)
Debris clearing	7	28	67
	(6.9)	(27.5)	(65.7)
Non-structural strategies			
Standalone plan	35	23	43
	(34.7)	(22.8)	(42.6)
Zoning	45	16	42
	(43.7)	(15.5)	(40.8)
Setbacks	34	28	42
	(32.7)	(26.9)	(40.4)
Protected areas	41	31	31
	(39.8)	(30.1)	(30.1)
Land acquisition	45	38	20
	(43.7)	(36.9)	(19.4)
Education	9	62	33
	(8.7)	(59.6)	(31.7)
Training	12	64	28
	(11.5)	(61.5)	(26.9)
Intergovernmental agreements	20	53	30
	(19.4)	(51.5)	(29.1)
Referendum	86	8	6
	(86.0)	(8.0)	(6.0)
Computer models	34	37	30
	(33.7)	(36.6)	(29.7)
Community block grants	52	41	9
	(51.0)	(40.2)	(8.8)
Construction codes	28	9	64
	(27.7)	(8.9)	(63.4)
Specific policies	18	26	59
	(17.5)	(25.2)	(57.3)
Land development codes	22	12	70
	(21.2)	(11.5)	(67.3)

[a] Row percentages are in parentheses.
Reproduced from Brody *et al.* (2009a) "Evaluating local flood mitigation strategies in Texas and Florida," *Built Environment*, **35** (4), 492–515, with permission.

Education, outreach, and training programs for residents and developers are also frequently used to reduce the adverse impacts of floods. For example, over 91% of communities use, either moderately or extensively, educational outreach efforts to inform households and businesses about the nature of local flood risks. Surprisingly, intergovernmental agreements are a regularly used technique, with over 80% of the sample coordinating their flood mitigation efforts with other authorities (although only 29% use this mechanism extensively). This result suggests broad recognition that collaboration across agencies and jurisdictional lines is necessary to address flood-related problems.

On the opposite side of the spectrum, the least used non-structural flood mitigation technique among respondents is the referendum, with only 14% of surveyed localities implementing this strategy. Local jurisdictions in Texas and Florida generally believe that flood control is the responsibility of technical experts and is not an issue for the general public to address. Despite the availability of federal funding for localities totaling approximately $116 billion since 1974, community development block grants through the housing and urban development department also receive low scores among non-structural mitigation techniques. Another infrequently used mitigation technique is the acquisition of land parcels to protect areas vulnerable to flooding. Local government purchase of land is often cost-prohibitive and controversial since it permanently removes from the tax base parcels that could potentially be developed in the future. However, land acquisition programs at the state level have been quite successful, particularly in Florida, which has several programs in place to acquire sensitive lands. The Preservation 2000 Initiative and the Florida Forever program, for example, used a documentary stamp tax to generate $300 million annually for acquisition of conservation lands (Beatley, 2000: 5–20). At the local level, Pinellas County adopted the Penny for Pinellas program requiring a 1-cent local sales tax, which applies to all sales, use, services, rentals, admissions, and other authorized transactions (Brody, 2008). Proceeds from the local option sales tax can be used only for capital projects. Of this money, approximately $3.3 million had been raised for parks and land acquisition by 2003 (www.pinellascounty.org/Penny/default.htm).

Examining the degree to which mitigation strategies are being implemented for the entire study area shows broad trends at the local level, but could mask important differences across the two states. Based on independent two-sample tests (Mann–Whitney U tests), we found that Florida uses retention and detention devices significantly more than Texas ($z = -2.443$, $p<0.05$). Landscaped retention and detention ponds are frequently integrated into the open spaces and common areas of suburban developments throughout Florida. Clearing of debris from drainage channels is also used by localities significantly more in Florida ($z = -2.455$, $p<0.05$), possibly because this state contains a greater number of urbanized areas where blockages are more likely to be noticed by local planning officials. Local jurisdictions in Florida also tend to have more financial resources to respond quickly to these potential flood problems.

Differences between the study states are even more pronounced for non-structural mitigation strategies. Florida's mandate that every local jurisdiction must adopt a legally binding comprehensive plan most likely drives their significantly higher implementation of land use planning tools to address local flooding, including: zoning ($z = -3.779$, p< 0.01), development setbacks ($z = -3.643$, p< 0.01), protected areas ($z = -4.785$, p<0.01), land acquisition ($z = -1.969$, p<0.05), and land development codes ($z = -4.974$, p<0.01). Florida localities are also more likely to use construction and building codes to reduce the adverse impacts of floods ($z = -2.726$, p<0.01). Localities in Florida are required to adhere to a stringent statewide building code adopted in 2002, which has become the gold standard in terms of reducing the adverse impacts of hurricanes and associated flooding on structures (Brody *et al.*, 2009a: 492–515).

Organizational capacity to mitigate flood damage

As previously discussed, the capacity of local organizations to adopt and implement flood mitigation strategies is an essential component to understanding the degree to which communities address flood problems. Overall, localities in Florida and Texas reported a generally strong commitment to developing flood-resilient communities (see Table 7.3). Over 70% of responding localities considered this characteristic to be strong or very strong within their organizations. The level of leadership is also an important organizational component to addressing floods: over 22% of respondents listed it as very strong. Following the same trend, verbal communication and information sharing among staff is highly rated. For example, verbal communication, which is an important part of building effective public organizations, was ranked as either strong or very strong by almost 70% of responding localities. In contrast, local jurisdictions in Texas and Florida reported low availability of financial resources to plan effectively for a flood-resilient community. Over 28% of respondents listed this capacity trait as weak or very weak, and only 5.4% considered the amount of financial resources within their organization to be very strong. Local planners and administrators also believe the number of staff members dedicated to flood mitigation is inadequate, as evidenced by less than 5% listing this attribute as very strong. Finally, survey respondents reported a low degree of public participation in the flood planning process, possibly because mitigation is often considered the domain for technical experts rather than for the general populace.

Examining differences in the strength of organizational capacity between Texas and Florida provides even more information about which state is better prepared

Planning decisions and flood attenuation

Table 7.3 *Frequency and descriptive statistics of organizational capacity for flood mitigation[a]*

Organizational capacity	Not present	Very weak	Weak	Neither	Strong	Very strong
Commitment	5	4	7	28	74	31
	(2.3)	(2.7)	(4.7)	(18.8)	(49.7)	(20.8)
Public officials	5	7	6	37	68	26
	(2.3)	(3.2)	(2.8)	(17.1)	(31.3)	(12)
Sharing information	3	4	6	36	69	31
	(2.0)	(2.7)	(4.0)	(24.2)	(46.3)	(20.8)
Verbal communication	3	2	5	35	77	27
	(2.0)	(1.3)	(3.4)	(23.5)	(51.7)	(18.1)
Sharing resources	9	3	8	59	49	20
	(6.1)	(2.0)	(5.4)	(39.9)	(33.1)	(13.5)
Networks	5	5	12	48	56	23
	(3.4)	(3.4)	(8.1)	(32.2)	(37.6)	(15.4)
Leadership	2	3	3	44	64	33
	(1.3)	(2.0)	(2.0)	(29.5)	(43.0)	(22.1)
Financial resources	7	18	24	50	42	8
	(4.7)	(12.1)	(16.1)	(33.6)	(28.2)	(5.4)
Available staff	3	12	28	48	50	7
	(2.0)	(10.1)	(29.1)	(32.4)	(33.8)	(4.7)
Data quality	4	9	16	44	53	21
	(2.7)	(6.1)	(10.9)	(29.9)	(36.1)	(14.3)
Public participation	10	10	24	60	37	6
	(6.8)	(6.8)	(16.3)	(40.8)	(25.2)	(4.1)
Adjustable policies	4	5	10	54	60	13
	(2.7)	(3.4)	(6.8)	(37.0)	(41.1)	
Long-range planning	6	6	19	44	55	16
	(4.1)	(4.1)	(13.0)	(30.1)	(37.7)	(11.0)
Human ecology	9	12	11	47	55	12
	(6.2)	(8.2)	(7.5)	(32.2)	(37.7)	(8.2)
Hire and retain staff	10	11	25	40	41	18
	(6.9)	(7.6)	(17.2)	(27.6)	(28.3)	(12.4)

[a] Row percentages are in parentheses.
Reproduced from Brody *et al.* (2009a) "Evaluating local flood mitigation strategies in Texas and Florida," *Built Environment*, **35** (4), 492–515, with permission.

to deal with local flooding. Overall, Florida localities scored higher than those in Texas on all 15 measures of organizational capacity. Specifically, results show that Florida localities garner significantly more interest from elected public officials when it comes to flood planning ($z = -2.339$, $p<0.05$). Commitment from elected officials is often the key trigger for implementation of suggested flood programs. Florida localities also seem to be significantly more capable of hiring and retaining key staff members over the long term ($z = -2.122$, $p<0.05$), which allows organizational knowledge and expertise to be maintained from one flood event to the next. The transfer of local knowledge across changes in management personnel is essential when dealing with a highly contextualized issue such as flooding.

The degree of financial resources committed to flood planning is another important organizational trait that is significantly higher in Florida than coastal Texas, as indicated by measure of annual budget ($z = -1.985$, p<0.05). Higher planning budgets enable localities to hire qualified staff and implement more costly yet effective strategies, such as land acquisition or the establishment of protected areas. Finally, local communities in Florida appear to have significantly more public participation throughout their flood planning processes ($z = -2.238$, p<0.05), most likely stemming from a state requirement and an active public. Engaging the public in decision making is often seen as essential to ensuring that adopted policies are implemented at the household level.

In the final analytical phase, we examined differences among Texas and Florida localities by combining individual indicators into indices of mitigation techniques and organizational capacity. We calculated depth and breadth measures to better gauge the overall differences associated with flood mitigation across the two states. As shown in Table 7.4, independent two sample t-tests show that localities in Florida score significantly higher than those in Texas on both depth and breadth of overall mitigation efforts ($t = -4.26$, p = 0.000; t = -3.24, p = 0.002).

While the structural mitigation indices are not statistically significant ($t = -1.10$, p = 0.272; t = 0.026, p = 0.979), Florida localities do score significantly higher for both breadth ($t = -3.93$, p = 0.000) and depth ($t = -4.64$, p = 0.000) of non-structural mitigation techniques. Perhaps most importantly, local jurisdictions in Florida rank significantly higher on our summary index of organizational capacity (47.59 versus 13.73 in Texas, $t = -2.35$, p = 0.020).

Our survey is one of the first to illustrate the status and trends associated with flood mitigation techniques at the local level along the Gulf of Mexico coast. Descriptive statistical results clearly show that localities in Florida implement, on average, more diverse and extensive flood mitigation policies. Strong state requirements for comprehensive planning that include flood mitigation, building codes, and public participation during the planning process most likely lend strength to city and county flood programs (see for example, Berke and French, 1994: 237–250; Berke *et al.*, 1996: 79–96; Burby, 2003: 33–49, 2005: 67–81; Burby and Dalton, 1994: 229–238; Burby *et al.*, 1997;). Top-down regulatory mechanisms are often controversial, but in this instance they hold local jurisdictions to a higher standard of mitigation and resiliency that can buffer against the adverse affects of flood hazards.

A much stronger organizational capacity to mitigate the adverse impacts of floods among Florida communities has most likely helped garner the commitment, expertise, and financial resources to implement a cadre of mitigation techniques. The adoption of legally binding comprehensive plans and land development codes

Planning decisions and flood attenuation

Table 7.4 *Mean comparisons of flood mitigation and organizational capacity indices*[a]

	Texas	Florida	Mean difference	t-test	p-value
Depth of mitigation	15.06	20.43	−5.37	−4.26	0.000
	(6.36)	(6.43)			
Breadth of mitigation	10.33	12.68	−2.35	−3.24	.002
	(3.53)	(3.79)			
Depth of structural	3.67	4.09	−0.42	−1.10	0.272
mitigation	(1.83)	(2.04)			
Breadth of structural	2.42	2.41	0.006	0.026	0.979
mitigation	(1.05)	(1.22)			
Depth of non-structural	11.40	16.34	−4.94	−4.64	0.000
mitigation	(5.44)	(5.38)			
Breadth of	7.92	10.27	−2.35	−3.93	0.000
non-structural	(3.09)	(2.99)			
mitigation					
Depth of organizational	13.73	47.59	−4.83	−2.35	0.020
capacity	(10.14)	(52.42)			

[a] Standard deviations are in parentheses. Based on F-test results, equal variances are assumed for all variables except depth of organizational capacity.
Adapted from Brody *et al.* (2009a) 'Evaluating local flood mitigation strategies in Texas and Florida', *Built Environment*, **35** (4), 492–515, with permission.

that address flooding and severe storms, acquisition of vulnerable areas, and education campaigns are just a few of the tools communities in Florida are using to combat chronic flooding. While these strategies can be expensive, time-consuming, and controversial, they may be more effective in reducing property damage and human losses from floods over the long term. Finally, the fact that Florida localities have a greater focus on public participation during the flood planning process may lead to greater community support and stronger implementation of adopted strategies (Brody, 2003a: 107–119). Overall, local public officials and residents in Florida appear more engaged in proactive planning for floods and less tolerant of a continual cycle of property damage and human casualties resulting from repetitive flooding. Consequently, more financial and human resources are being spent on implementing local flood programs in Florida.

Factors driving flood mitigation strategies: the role of organizational capacity

The capacity of an organization is so essential to developing and implementing flood mitigation programs that the issue deserves special attention. Correlation analysis

Table 7.5 *Correlations between organizational capacity characteristics and mitigation strategies*

	Structural mitigation	Non-structural mitigation
Organizational capacity	0.32**	0.45**
Commitment	0.19	0.41**
Sharing information	0.23*	0.40**
Verbal communication	0.39**	0.32**
Sharing resources	0.07	0.30**
Networks	0.14	0.30**
Leadership	0.31**	0.28**
Financial resources	0.27**	0.26*
Available staff	0.30**	0.34**
Data quality	0.24*	0.34**
Adjustable policies	0.34**	0.27**
Long-range planning	0.35**	0.36**
Human ecology	0.39**	0.48**
Hire and retain staff	0.26*	0.32**
Commitment of public officials	0.20	0.40**
Public participation	0.26*	0.34**

*$p<0.05$, **$p<0.01$.
Adapted from Brody *et al.* (2009b) "Identifying factors influencing flood mitigation at the local level in Texas and Florida: the role of organizational capacity," *Natural Hazards*, **52** (1), 167–184, with permission.

(shown in Table 7.5) further demonstrates the strong statistical link between organizational capacity and the implementation of both structural and non-structural mitigation strategies (Brody *et al.*, 2009b: 167–184). In general, a high degree of local organizational capacity is more of a catalyst for the adoption of non-structural techniques, which often require a greater degree of collaboration among multiple parties for effective implementation. For example, staff commitment to planning for flood-resilient communities is significantly correlated with non-structural ($p<0.01$), but not structural, approaches, such as engineering-based interventions. The different effect of organizational capacity on structural versus non-structural flood mitigation strategies is particularly evident in the sharing of financial and personnel resources among staff members (in the same organization and in other organizations within the jurisdiction). While structural solutions are not significantly correlated (since presumably these initiatives come from one organizational source), non-structural activities are highly significant ($p<0.01$) because multiple parties are often involved in implementing policy-related techniques. Overall, building relationships is important because of the process-oriented, collaborative requirements of non-structural flood mitigation. The establishment of informal

Table 7.6 *Modeling structural flood mitigation strategies*

Variable	Coefficient	Standard error	t-value	Significance	Beta
Organizational capacity	0.0446	0.0155	2.87	0.005	0.3070
State (FL=1)	−0.5353	0.4264	−1.26	0.213	−0.1367
Floodplain (%)	−0.0285	0.0093	−3.05	0.003	−0.3261
Education	0.1740	0.3480	0.50	0.618	0.0525
Five-year flood loss	0.0000	0.0000	0.29	0.770	0.0297
Recent flood event	−0.1668	0.0878	−1.90	0.061	−0.1956
Constant	2.7207	0.8790	3.10	0.003	
R2	0.2060				
n	88				

Adapted from Brody *et al.* (2009b) "Identifying factors influencing flood mitigation at the local level in Texas and Florida: the role of organizational capacity," *Natural Hazards*, **52** (1), 167–184, with permission.

or personal networks among staff members follows a similar statistical pattern, where this indicator is statistically significant for non-structural (p<0.05), but not for structural, techniques.

Even when an indicator of organizational capacity is significantly associated with flood mitigation, the correlation is almost always stronger for non-structural strategies. For example, the correlations for information sharing among staff members, data quality, commitment of public officials, and public participation are all one statistical magnitude stronger for non-structural activities. The one characteristic of organizational capacity that increases the implementation of structural mitigation is available financial resources to develop flood-resilient communities. Indeed, engineering-based solutions are usually more expensive than those efforts rooted in planning and education, thus requiring greater amounts of financial resources to implement.

Organizational capacity, in its index form, remains a statistically significant predictor of increased flood mitigation strategies in Texas and Florida, even when controlling for various contextual variables (Brody *et al.*, 2009b). As shown in Table 7.6, a unit increase in the overall organizational capacity measure results in a significant increase in the extent to which structural measures are implemented to reduce the adverse impacts of floods (p<0.01). Interestingly, within the same model, more recent damaging flood events, rather than the total amount of property loss reported over the entire study period, correspond to a higher use of structural techniques (p<0.1).

Our organizational capacity index has an equally strong effect on the implementation of non-structural mitigation strategies (Table 7.7). In this model, the state in

Table 7.7 *Modeling non-structural flood mitigation strategies*

Variable	Coefficient	Standard error	t-value	Significance	Beta
Organizational capacity	0.1208	0.0413	2.92	0.004	0.2830
State (FL=1)	–3.7574	1.1370	–3.30	0.001	–0.3268
Floodplain (%)	–0.0235	0.0208	–1.13	0.261	–0.0916
Education	3.3210	0.8831	3.76	0.000	0.3415
Five-year flood loss	0.0000	0.0000	2.52	0.014	0.1584
Recent flood event	0.0393	0.1348	0.29	0.771	0.0157
Constant	6.2143	2.3905	2.60	0.011	
R2	0.4096				
n	88				

Adapted from Brody *et al.* (2009b) "Identifying factors influencing flood mitigation at the local level in Texas and Florida: the role of organizational capacity," *Natural Hazards*, **52** (1), 167–184, with permission.

which these initiatives are being implemented is once again statistically relevant. As discussed above, Florida uses significantly more non-structural measures to mitigate floods compared with Texas, which relies more heavily on engineering-based approaches. Education level within communities is also a notable driver of non-structural mitigation techniques (p<0.01). An educated public will most likely be more receptive to strategies that entail information dissemination, training, and land use planning projects. In fact (based on the reported standardized betas), education is the strongest predictor of non-structural mitigation techniques in our model, more so than the two flood history variables combined (Brody *et al.*, 2009c: 912–929). In contrast to structural mitigation, in this model, the most recent year of a damaging flood event has little effect on the adoption of non-structural mitigation techniques. Instead, total losses from floods over the previous five years are a significant predictor (p<0.05) of increased usage of non-structural mitigation strategies.

Summary of results

Based on the results above, the degree to which an organization responsible for flood management has the resources, expertise, and culture in place to mitigate the negative effects of floods is an essential factor in shaping the type of flood mitigation program at the local level. In fact, organizational capacity is as or more important than past disaster experience, geophysical conditions, and the state in which planning is taking place. Given the importance of collaboration, expertise, financial resources, and other characteristics comprising organizational capacity, building capable organizations should be a priority for decision makers whose goal is to protect their communities from flood-related disasters.

In addition to showing the criticality of organizational capacity in facilitating the development of resilient communities in Texas and Florida, our analysis also uncovers other factors driving the implementation of local flood mitigation strategies. Specifically, the state's political and regulatory climate is a major factor leading to more extensive implementation of non-structural activities. As already mentioned, Florida implements significantly more non-structural techniques most likely because it has a much stronger planning tradition where, by mandate, localities must adopt a comprehensive plan that addresses flooding issues. Land use planning often involves zoning, land acquisition, protected areas, education, and other activities considered to be non-structural approaches to flood mitigation. It is also apparent that local officials and the general public in Florida are more engaged in proactive planning for floods and less tolerant of property damage and human casualties resulting from flood events, leading localities to implement a more extensive array of mitigation strategies. A greater commitment to planning and preparedness at the state level may thus result in more extensive mitigation strategies at the local level, resulting in decreased property damage and human casualties resulting from floods.

We also find that the education level of residents within a community is a strong driver of non-structural flood mitigation techniques. We posit that an educated public is more likely to be receptive to targeted education, training, and referendums, comprehensive plans, and other non-structural flood mitigation strategies. They may also be more cognizant of the long-term benefits of non-structural approaches and the previous failure of structural approaches highlighted in the media during hurricanes and large tropical storms in both states.

Finally, our analysis shows that prior experience with floods affects the implementation of flood mitigation strategies in different ways. For example, recent damaging flood events appear to activate structural mitigation techniques, while a history of repetitive flood loss triggers the adoption of non-structural techniques (Brody *et al.*, 2009b: 167–184). This difference can be explained through the timing of local organizational response to flood events. The implementation of structural approaches often occurs as a reaction to a single flood event. For example, clearing of debris, channelization, and hardening initiatives are often quick responses that require little public input or long-term planning. In contrast, a longer-term history of flood damage will possibly prompt the need for non-structural strategies requiring more time and public commitment to implement. In general, non-structural approaches to mitigation tend to be focused more on behavioral changes over the long term rather than accruing quick gains in response to a single event. Understanding the chronic nature of coastal flooding along the Gulf coast will be critical to implementing policies that sustainably shape the way communities develop over the long run.

Mitigation techniques and flood damage

While understanding the factors contributing to the implementation of flood mitigation strategies is important for developing strong flood reduction programs at the local level, it does not address the question central to this book: do these techniques actually reduce the adverse impact of floods? Based on the above analyses, we know that using FEMA's CRS scores as a proxy for local flood mitigation, Florida is approximately twice as prepared as coastal Texas to address flood-related problems. We also note that Florida has adopted significantly more mitigation techniques, experiences more yearly precipitation, has built more expensive structures in areas vulnerable to flooding, and has a larger percentage of its population living within the 100-year floodplain. But, over the five years preceding the release of our survey, Texas recorded significantly higher property damage per person from floods and more than twice the number of human casualties from storm events.

To move further in our understanding of the effectiveness of flood mitigation, we need to empirically examine the relationship between specific techniques and actual flood loss. To accomplish this goal, we matched each technique examined above with reported per capita flood loss based on NFIP claims for the two-year period following the survey. Through this analytical approach, we can test the relationship between the implementation of various flood mitigation strategies and observed flood losses over time. Table 7.8 begins with mitigation and organizational capacity techniques in an index form. Based on Pearson's product moment statistical correlations, non-structural techniques appear to significantly reduce property damage from floods across the study area, while structural techniques have a negative but nonsignificant impact.

Examining the relationship between specific mitigation techniques and reported flood losses helps identify which approaches may be most beneficial for local policy makers to adopt. Table 7.9 shows that, of all the structural approaches assessed, only storm-water retention or detention ponds are significantly related to reductions in insured property damage from floods (and only mildly). Surprisingly, dams have a negative, but nonsignificant, association with property damages from floods. These results are consistent with parallel analyses that control for various contextual variables. For example, at the watershed level, dams have a weak impact on reducing the amount of flooding across both states (Brody *et al.*, 2007b: 413–428). When assessing property damage from floods in Florida, a county's CRS rating

Table 7.8 *Overall mitigation and flood damage*

	Flood damage
Structural	−0.0777
	0.4401
Non-structural	−0.2711**
	0.0061
Mitigation	−0.2563**
	0.0097
State	0.5674**
	0.0000

*p<0.1, **p<0.05.

is more than twice as effective as the presence of dams in reducing flood damage (Brody *et al.*, 2007a: 330–345). Only for high-impact floods do dams become effective at reducing the adverse consequences of floods. Even in this situation, it would take 29 dams to decrease the odds of a high damage flood by only 22.6%.

To further put the effectiveness of dams into perspective, a two-class jump in CRS rating for a county in Florida would provide the same amount of reduction in flood damages without the expense and risk of structural failure. Our sister study in Texas using similar data shows that wetland protection may be more effective than dams in mitigating property loss over time. In dollar terms, the presence of a dam resulted in a $27 290 decrease in the average property damage for each flood event across the 37 counties in the sample. This may seem worthwhile, except that based on our model, only 129 wetland alteration permits offset the flood-reducing effects of dams. Dams also have several undesirable side effects that wetland protection does not: they are extremely costly mitigation alternatives; they can intensify development in flood-prone areas out of a false sense of security; they can present a hazard in the case of structural failure; and these structures tend to be politically contentious (Pielke, 1999).

In contrast, almost half of the non-structural techniques listed in our survey are significantly related to a reduction in NFIP-reported losses from flood events (Table 7.9). Having a specific flood policy contained within a comprehensive plan has the strongest statistical correlation with damage reduction. This result indicates the effectiveness of comprehensive planning, which usually serves as a spatial blueprint for future development patterns. Comprehensive plans consider a broad range of growth and development issues so that a specific flood policy would be embedded into the overall future land use pattern of a community, where it will most likely be more effective in attaining intended mitigation goals. Comprehensive plans may also be

Table 7.9 *Specific mitigation techniques and flood damage*

	Flood damage[a]
Structural strategies	
Retention/detention	−0.1498*
Levees	0.1557
Channelization	0.0624
Dams	−0.0717
Debris clearing	−0.1052
Non-structural strategies	
Stand-alone plan	0.0781
Zoning	−0.1893**
Setbacks	−0.2737**
Protected areas	−0.3457**
Land acquisition	−0.1588*
Education	−0.0996
Training	0.0387
Intergovernmental agreements	0.1207
Referendum	−0.0417
Computer models	−0.1177
Community block grants	−0.0438
Construction codes	−0.2719**
Specific policies	−0.3718**
Land development codes	−0.3158**

[a] Per capita logged NFIP loss estimates from 2006 to 2007.
*$p<0.1$, **$p<0.05$.

the most likely instrument to guide growth away from areas vulnerable to flooding because they tend to set the regulatory framework for a locality after adoption. It is interesting to note that, by contrast, stand-alone flood plans, by themselves, have no significant connection to observed decreases in flood losses. These types of plan are often nonbinding or are integrated into local land use and development-based decisions. While they are more focused and detailed policy instruments, they often do not have the political or legal backing to be implemented at the local level.

Land development code regulations specific to flood mitigation also appear to associate with a local reduction in flood losses. The more extensively localities use this regulatory vehicle, the lower the observed flood damage. The use of protected areas to prevent development in flood-prone locations is also significantly correlated with a reduction in property damage. Establishing protected areas essentially removes the threat of damaging storm events within that particular designation. In other words, if there are no structures in flood-prone areas, there will be no damages to report.

Similarly, instituting development setbacks or buffers adjacent to riverine areas or floodplains is negatively associated with property damage caused by floods. This technique also prevents structures from being situated in highly vulnerable areas. Zoning provisions, where specific areas are designated for a particular use, is another spatially oriented flood mitigation strategy that significantly reduces economic impact. If done correctly, vulnerable or sensitive areas are designated for low-intensity development, or even conservation, to minimize the adverse impacts of flooding and long-term economic disruption for a community. Finally, local construction codes that address flooding correspond to a significant reduction in observed flood damage. If locating structures in vulnerable areas cannot be prevented, ensuring their construction is resistant to potential flooding impacts is a second-line approach to mitigation at the local level.

In general, non-structural mitigation strategies appear to be more effective in reducing flood damages from both rainfall and surge-based events. Specifically, land use strategies that target vulnerable areas significantly reduce insured property losses. Set-backs, protected areas, and land acquisition techniques accomplish mitigation goals by helping to remove human activities from the most sensitive areas.

Based on these statistical clues, a spatially targeted flood mitigation program may be the best way to protect coastal communities rather than casting a broad regulatory net. It is also important to note that while non-structural strategies appear to be more effective in reducing observed flood damage, highly resilient communities will most likely rely on a hybrid approach that combines both structural and non-structural mitigation techniques. Since every community has its own set of contextual characteristics, a complementary mix of flood mitigation strategies can produce a synergistic effect of reduced losses from repetitive flooding. Decision makers should not only take clues from evidence-based research but also devise a well-balanced, integrated flood program that caters to the specific needs of their community.

Organizational capacity is also a potentially important contextual aspect of reducing flood impacts (Table 7.10). While the overall index is essential for the adoption of flood mitigation strategies, by itself it does not correlate with a marked reduction in observed damage. However, several individual characteristics are important to reducing insured flood damage. First, interest from elected public officials in planning for a flood-resilient community appears to be a key element in implementing flood mitigation strategies. Many mitigation initiatives require approval from local elected officials or at least benefit from their support. Second, as expected, the amount of financial resources dedicated to flood planning leads to significant ($p<0.1$) reductions in insured damages. Allocation of funding allows decision makers to implement more extensive and far-reaching flood programs. Third, the number of staff available for flood planning and management is an

Table 7.10 *Organizational capacity characteristics and*
flood damage

	Flood damage[a]
Organizational capacity	−0.1311*
Commitment	−0.0654
Sharing information	−0.0519
Verbal communication	−0.0281
Sharing resources	−0.1173
Networks	−0.1144
Leadership	0.0491
Financial resources	−0.1564*
Available staff	−0.1725**
Data quality	0.0056
Adjustable policies	−0.0658
Long-range planning	−0.0717
Human ecology	−0.0584
Hire and retain staff	−0.1558*
Commitment of public officials	−0.1869**
Public participation	−0.0972

[a] Per capital logged NFIP loss estimates from 2006 to 2007.
*$p<0.1$, **$p<0.05$.

important correlate with less property damage from flooding. More personnel translate into a greater degree of expertise and effort directed at local flood problems. The number of staff dedicated to flood mitigation is a traditional measure of planning capacity shown to result in stronger outcomes, but should be considered only one aspect of a capable organization. Finally, equally important as the number of staff devoted to flood management is the ability to retain key staff members over the long term. The stability of personnel within an organization facilitates a more intimate knowledge of flood issues specific to a jurisdiction, as well as allowing this knowledge to transfer across different administrations.

Overall, organizational capacity is not associated with direct reductions in property damage as much as with the adoption of mitigation techniques. However, specific characteristics do appear important to decreasing flood impacts and can provide decision makers with tangible evidence on how to construct an organization that can effectively address flooding. Of course, this analysis should only be considered a starting point for understanding the relationships between mitigation, organizational capacity, and property damage from floods. More detailed statistical models that control for multiple contextual variables are warranted before any final conclusions can be made.

Conclusions

Overall, non-structural flood mitigation techniques can provide a powerful segue to developing more resilient communities along the Gulf coast. In this chapter, we show through empirical evidence and multiple analytical methods that the implementation of specific land use planning strategies significantly reduces the amount of flood damage incurred at the local level. Spatial management that targets the most vulnerable areas can allow development to occur, while at the same time lessening the adverse economic impacts of chronic flooding. Moreover, effective policies must be adopted and implemented by government organizations with the necessary capacity to ensure that flood reduction programs are binding and long-lasting. Proactive mitigation measures are thus a critical element for localities interested in establishing the linkages between sustainable planning and the reduction of floods.

Appendix 7.1 *Local flood mitigation survey instrument*

Flood policy response and planning capacity survey

Definitions
- Repetitive flooding occurs when the same physical location floods regularly or at a minimum of once per five years.
- Repetitive flooding can include, but is not limited to, structural damage.
- Flooding does not need to occur only as a result of major storms, but can take place even in response to relatively low amounts of precipitation.
- This type of flooding occurs chronically over time in the same general area.
- Flooding can result in structural damage, roadway damage, and disruption of hydrologic definition.

Purpose
- This survey seeks to understand how and why communities vary in their responses to localized repetitive flooding.

Instructions
Please answer the questions to the best of your ability.
You may need to consult with co-workers regarding some of these questions.

1. Over the last 5 years, how many floods have occurred in your jurisdiction? Circle the best response.
    0   1   2   5   6   10   10 or more
    If you responded 0, or no floods in the past 5 years, please skip to question 4.
    The next questions are about your jurisdiction's use of various techniques in response to a flood or floods.
2. Over the last 5 years, how often did your jurisdiction use the following *structural approaches* when responding to repetitive flooding? *For this survey, repetitive flooding occurs when the same physical location floods regularly or at a minimum of once per five years. Repetitive flooding can include, but is not limited to, structural damage.*

Please indicate the extent to which your jurisdiction used a response strategy by using the following scale:	Never used	Used occasionally	Used extensively	Not within this juris-diction's authority
a. retention/detention/holding	☐	☐	☐	☐
b. levees	☐	☐	☐	☐
c. channelization	☐	☐	☐	☐
d. dams	☐	☐	☐	☐
e. clearing of debris	☐	☐	☐	☐
f. other (please explain): _____	☐	☐	☐	☐

3. Over the last 5 years, how often did your jurisdiction use the following *non-structural or policy-related approaches* when responding to repetitive flooding?

Please indicate the extent to which your jurisdiction used a response strategy by using the following scale where:	Never used	Used occasion-ally	Used exten-sively	Not within this juris-diction's authority
a. Standalone flood plan	☐	☐	☐	☐
b. Zoning	☐	☐	☐	☐
c. Setbacks or buffers	☐	☐	☐	☐
d. Protected areas or conservation overlays	☐	☐	☐	☐
e. Land acquisition (e.g., fee simple purchase, purchase of development rights, conservation easements, etc.)	☐	☐	☐	☐
f. Education/outreach programs	☐	☐	☐	☐
g. Training/technical assistance	☐	☐	☐	☐
h. Intergovernmental agreements	☐	☐	☐	☐
i. Referendum (tax)	☐	☐	☐	☐
j. Computer models/evaluation systems (e.g., HEC)	☐	☐	☐	☐
k. Use of Community Development Block Grants (CDBG) to mitigate flooding problems	☐	☐	☐	☐
l. Construction codes	☐	☐	☐	☐
m. Specific policies in the local compre-hensive plan	☐	☐	☐	☐
n. Land Development Code regulation	☐	☐	☐	☐
o. other (please explain):	☐	☐	☐	☐

The next set of questions is about your jurisdiction's *ability to respond* to repetitive flooding events. There are many characteristics that help organizations adapt and effectively respond to repetitive flooding.

4. Over the last 5 years, how strong would you say the following characteristics have been in your jurisdiction's flood planning and/or hazard mitigation organization?

Please indicate the strength of each characteristic in your organization by using the following scale:	Not present	Very weak	Weak	Neither weak nor strong	Strong	Very strong
	☐	☐	☐	☐	☐	☐
a. commitment to planning for a flood-resilient community	☐	☐	☐	☐	☐	☐
b. interest from elected public officials in planning for a flood-resilient community	☐	☐	☐	☐	☐	☐
c. sharing of information among staff members (in the same organization or in other organizations within the jurisdiction)	☐	☐	☐	☐	☐	☐
d. verbal communication among staff members (in the same organization and in other organizations within the jurisdiction)	☐	☐	☐	☐	☐	☐
e. sharing financial and personnel resources among staff members (in the same organization and in other organizations within the jurisdiction)	☐	☐	☐	☐	☐	☐

Please indicate the strength of each characteristic in your organization by using the following scale:	Not present	Very weak	Weak	Neither weak nor strong	Strong	Very strong
	□	□	□	□	□	□
f. establishment of informal or personal networks among staff members (in the same organization and in other organizations within the jurisdiction)	□	□	□	□	□	□
g. degree of leadership in the organization's administration	□	□	□	□	□	□
h. available financial resources to plan effectively for a flood-resilient community	□	□	□	□	□	□
i. available staff members and other personnel to plan effectively for a flood-resilient community	□	□	□	□	□	□
j. quality of data (e.g., flood vulnerability, natural resources, GIS data layers, etc.) with which to plan effectively for a flood-resilient community	□	□	□	□	□	□
k. degree of public participation/ involvement in the planning process	□	□	□	□	□	□

Please indicate the strength of each characteristic in your organization by using the following scale:	Not present	Very weak	Weak	Neither weak nor strong	Strong	Very strong
	□	□	□	□	□	□
l.　ability to adjust policies in response to a flood-related problem (i.e., be flexible and adaptive in planning approaches)	□	□	□	□	□	□
m.　ability to think and plan long-range (20+ years)	□	□	□	□	□	□
n.　ability to make policies that recognize an interaction between human and ecological systems	□	□	□	□	□	□
o.　ability to hire and retain key staff members over the long term (i.e., personnel turnover rate)	□	□	□	□	□	□
p.　ability to adjust local policy in response to declining downstream water quality	□	□	□	□	□	□

The following questions will provide us with background information on your jurisdiction.

5. How many full-time professional staff members are dedicated to planning and flood mitigation in your jurisdiction? (*e.g., If you are the only person and split your time between 4 different roles evenly, put 0.25. If there are two full-time staff and one part-time staff persons, put 2.5*). _____ Full-Time Employees

6. Give an example of a recent flood you consider to be repetitive:
a. Date: Month: _____ Day: _____ Year: _____ b. Location (be as precise as possible):___

7. Estimate your organization's annual budget dedicated to flood planning: $0 – $5,000; $5,001 – $10,000; $10,000 – $20,000; $20,001 – $50,000; $50,001 – $100,000; $100,001 – $300,000; $300,001 or greater

8.How many years' experience do you have as a floodplain administrator? 0–1, 2–5, 6–10, 10 or greater years

9. How long have you worked for this organization? 0–1, 2–5, 6–10, 10 or greater years

10. Name of your jurisdiction (City or County name & State):

11. Your job title (e.g., "Floodplain Administrator" or "City planner"):

12. How many events with property damage have occurred in your local jurisdiction in the past 5 years? 0, 1, 2, 3, 4, 5, 6, 7, 8, 9, 10 or more

8

Other factors influencing flooding and flood damage

The bulk of our book focuses on the effects of wetland alteration and mitigation on flooding and associated flood damages. However, these factors can only be understood within the context of other variables influencing the problem of flooding. In Chapter 5, we identified many of these variables and described their expected influence on flooding and flood loss across the study area.

This chapter presents our empirical findings on the impacts of characteristics we previously treated as contextual control variables: natural environment, built environment, socioeconomic, and political/administrative. Within each category, we systematically discuss the effect of each variable in hopes of providing a broader understanding of the factors contributing to flood problems across the study area.

A distinguishing aspect of our results compared to other studies is that when we describe the individual effects of each contextual variable, we control for a multidisciplinary set of other factors affecting the nature of floods. We also draw conclusions based on multiple models addressing the flooding problem from different angles, thus offering the reader a more comprehensive view of the underlying causes and consequences of coastal flooding in Texas and Florida.

Natural environment

While human interventions can certainly reduce the adverse impacts of floods, their occurrence is largely driven by the natural environment. In most of our explanatory models, geophysical and climate-based factors explain approximately 30% of the variance for both flooding and related flood damage. As expected, precipitation is consistently the most statistically powerful predictor of flooding. In fact, the amount of precipitation is always the first or second most significant factor among all variables analyzed. For example, in an analysis of 85 watersheds across Texas and Florida, precipitation (the number of annual wet days) had by far the greatest influence on flooding, based on the number of times stream gauges exceeded their 12-year average from 1991 to 2002 (Brody *et al.*, 2007b: 413–428).

The same pattern holds when forecasting property loss from floods. Among all counties in Florida, average surface precipitation, recorded by county weather stations the day of and day before a flood event, was the strongest predictor of property damage (p<0.001) from 1997 to 2001. Furthermore, a unit increase in precipitation raised the probability of experiencing a high-damage flood event (above the median of $50 000) during this five-year period by almost 20% (Brody *et al.*, 2007a: 330–345). Even more telling, a standard deviation increase in precipitation of 4.2 inches increases the odds of a costly flood by approximately 111%.

In addition to rainfall intensity, our analyses identified two major characteristics associated with precipitation that influence floods: duration and timing. In both Texas and Florida, the duration of a storm event significantly increases (p<0.05) observed property damage. As one would expect, longer-lasting storms tend to produce more rainfall, exacerbating economic losses. Also, once the ground is saturated, additional precipitation may amplify adverse impacts due to ponding and increased sheet flow. From 1997 to 2001, in Florida, every additional day of rainfall during a storm event added an average of over $6640 in overall property damage per flood. During the same period in coastal Texas, any storm lasting more than one day created, on average, $182 007 of added damages for each recorded flood.

Related to the issue of timing is the total duration of a storm. For example, in Texas the amount of precipitation the day before the actual flood event is a stronger predictor of damage than that on the day of the event itself. In fact, heavy precipitation the day before the flood was by far the strongest determinant of total property damage. This finding is most likely a function of the delay between initial rainfall and resulting rise in water levels. As indicated above, saturated soil from rainfall can transform even modest amounts of precipitation during subsequent days into damaging flood events by reducing the absorption capacity of hydrologic systems. Even in urban areas where this lag time between initial rainfall and peak discharge is compressed by increased impervious surfaces, it is critical for decision makers and the public to understand that heavy precipitation, even when followed by sunny skies, may still result in significant flood damage the next day. An early response to the relatively slow onset of flood waters may help coastal communities avoid loss of property and human lives. Informing the public of this issue through educational programs could be even more effective in mitigating the adverse impacts of flood events.

Another noteworthy natural environment characteristic influencing the extent of flood damage is the amount of 100-year floodplain within a local community. A greater area or percentage of floodplain is generally associated with increased storm damage because there is a higher likelihood that development will occur in vulnerable areas, leading to greater economic losses. However, the influence of this variable is not as statistically strong as expected. We offer several explanations for this weakened effect: local decision makers are taking precautions to avoid development in floodplains, thus reducing the amount of damage floods may cause; development that does occur in the floodplain is accompanied by stringent mitigation techniques through the CRS or other flood programs; or the natural boundaries of the floodplain have been so altered by filling, draining, grading, and channelization that this measure is no longer a meaningful predictor of property damage incurred from local storm events. In any case, having more floodplain in a local community does not automatically trigger significantly higher amounts of flood damage over time.

Human and built environment

In addition to the human and built environment interventions already discussed in Chapter 6 and Chapter 7, several other variables that impact flooding and associated damage should be noted. First, the amount of impervious surfaces within a community has long been identified as a major human-induced environmental change contributing to increased flood-related impacts. However, when we tease out the alteration of naturally occurring wetlands (see Chapter 6) from simply the amount of impervious surface itself, the effect is quite diminished (Brody *et al.*, 2007b: 330–345, 413–428, 2008b: 1–18). The statistical effect of impervious surfaces is only marginally significant when a wetland alteration variable is also included, in almost every model we analyze (the correlations between wetland alteration and impervious surface variables were not high enough to pose a multicolinearity problem for any of the models). Notably, imperviousness does not necessarily require the loss of wetlands, depending on where the surface is located. By separating wetland loss from amount of impervious surfaces, we lessen its statistical effect by removing what is possibly the most essential adverse consequence of impervious surfaces: loss of wetlands (for more details, see Chapter 6).

This finding has important implications for watershed planners, floodplain administrators, and local decision makers. While halting development may not always be possible, regulating both the degree to which a watershed is converted to impervious surface and the surface's location may be key to reducing the impact of flooding events. Concentrating buildings, parking lots, roads, and other hardened surfaces in the least vulnerable areas (e.g. away from wetlands and outside the 100-year floodplain) can help retain the functionality of existing hydrological systems and reduce the severity of flooding over the long term.

In addition to the amount and location of impervious surfaces is the concept of density. Density refers to the number of people or structures in a given area and can be a useful measure of form or overall pattern of the built environment. Interestingly, the degree of development density can have contrasting consequences when it come to flooding and flood impacts. Most anti-sprawl and new urbanist advocates support high-density forms of development as being more sustainable and hazard-resilient. However, if high density occurs in areas of vulnerability to floods, such as low-lying areas that receive large amounts of rainfall, more people and property will be exposed to flooding (see Berke *et al.*, 2009: 441–455; Stevens *et al.*, 2009: 605–629). Conversely, low-density development patterns could reduce the overall impacts of floods by spreading out human settlements across flood-vulnerable landscapes. On the other hand, low-density per capita population is an indicator of sprawl associated with increased impervious surfaces and fragmentation of hydrological systems (Brody *et al.*, 2006b: 75–96). Low-density development may also place a greater number of people and structures in harm's way.

Of course, measures of density are so reliant on the area being considered that we may become "prisoners of perspective." For example, population density as calculated for large watersheds may show different effects on flood damage than if the same area was measured at the Census Block Group level. The first spatial perspective could show an overall low-density pattern of development while the second could show high density. Indeed, we found this phenomenon to be true in our own analyses. Population density calculated for 85 watersheds within the coastal margin of Texas and across all of Florida based on the USGS fourth-order hydrological unit code had no effect on observed flooding (Brody *et al.*, 2007b: 413–428). In contrast, higher population density, as measured by the U.S. Census, at the county level in Florida correlated significantly ($p<0.01$) with increased property damages from floods between 1997 and 2002 (Highfield and Brody, 2006: 23–30). Among 74 counties in eastern Texas, higher population density also significantly increased observed human casualties from floods during a similar period (Zahran *et al.*, 2008: 537–560).

While the scale debate is important from a statistical perspective, the fact is that land use and development decisions are made at the county and city administrative levels. Ecosystems, watershed units, census block groups, and other areas of aggregation (that may be more appropriate planning units) do not have decision-making authority. If development density is important to control with the goal of reducing flood losses over time, it must be done at the appropriate administrative level where public decision makers and local elected officials are organized.

Socioeconomic factors

Our research has also shown that the socioeconomic composition of communities in Texas and Florida is an important predictor of flooding and flood losses over time. Measures of wealth and income, in particular, are highly correlated with flooding and flood impacts and have been overlooked in the past by physical science researchers modeling the severity of storms. Wealthy communities across the study area are both advantaged because they have more resources with which to combat floods and disadvantaged because more expensive structures are often built in vulnerable locations. How we choose to measure wealth and the flood problem, then, will determine their observed relationships.

For example, building expensive "second homes" in a floodplain will most likely result in a greater amount of property damage and associated losses when storms hit. Indeed, in Florida high housing improvement and density of housing values significantly increased property damage incurred from floods (Brody *et al.*, 2007a: 330–345; Highfield and Brody, 2006: 23–30). In this sense, wealth is associated with greater losses. However, when wealth is measured by the income of residents, we see different results. For example, when predicting flooding at the watershed level based on observed stream gauge data in both study states, median household income has a negative statistical effect. That is, on average, wealthier residents living in coastal areas experience a lower number of flooding events (Brody *et al.*, 2007b: 413–428). We suspect that local communities with the financial resources, leadership, and planning expertise to implement both structural and non-structural (e.g., land acquisition, zoning, education programs) mitigation strategies are typically better protected from the threats of persistent floods. Past research has shown that wealthier communities adopt higher-quality plans with respect to mitigating natural hazards such as floods (Berke *et al.*, 1996: 79–96), and our own analyses confirm this outcome. As shown in Chapter 7, localities with greater financial resources are significantly more likely to implement flood mitigation techniques, particularly structural solutions that require higher capital investments (Brody *et al.*, 2009b: 167–184). This result may also contribute to the explanation of why Florida, which is wealthier than coastal Texas, has a comparatively lower number of flood events.

Does this mean that lower income or more socially disadvantaged communities are more at risk from flooding because they lack the financial capacity to reduce the threat? In eastern Texas, Zahran *et al.*, 2008, examined the relationship between socially vulnerable communities measured primarily through family income and human casualties from flood events (death and injuries combined) from 1997 to 2001. The results were revealing: a unit increase in the measure of disadvantaged populations increased the odds of death or injury from a flood event by 42.4%. Counties with above-average compositions of poor residents are significantly ($p<0.01$) more likely to experience human injury and deaths from flooding, even when controlling for natural and built environment characteristics such as precipitation and impervious surfaces. This finding empirically demonstrates that lower income communities suffer disproportional adverse impacts from floods because, in part, they lack the financial resources to address the problem with appropriate mitigation techniques.

Political/administrative

A final category of variables that can influence the number and extent of floods is the specific political and administrative context within which localities operate. States in particular have their own set of mandates, institutional frameworks, political agendas, and commitment to mitigating the adverse impacts of floods. Localities also have their own agendas and policies, but often take cues from the state level. For example, in Chapter 5, we indicated that, based on previous research (see Berke and French, 1994: 237–250; Burby, 2005: 67–81; Burby *et al.*, 1997, among others), state planning mandates increase the quality and effectiveness of plans with regard to floods and other natural hazards. Our study corroborates these past findings when examining a state with a mandate to plan (Florida) and another without (Texas). As shown in Chapter 7, Florida has implemented a significantly stronger set of flood mitigation strategies. For almost every mitigation technique we analyzed, Florida emerges the winner hands-down. Along with its mandate comes a higher degree of local organizational capacity to make the policies stick over the long term.

The question remains of whether all of the structural and non-structural mitigation measures implemented in Florida make a difference in terms of the degree of impact jurisdictions experience from flooding. The empirical answer is unmistakable: in every bi-state model we analyzed stretching back 20 years, coastal Texas has considerably more flooding events, property damage, and human casualties from floods than does Florida. This finding may at first seem counterintuitive since Florida has more wetland alteration permits (see Chapter 6), greater impervious surface coverage, more expensive structures in vulnerable areas, and higher population growth rates. Also, rainfall, the strongest predictor of flooding and flood damages, is much greater in Florida than in coastal Texas. Statistically controlling for these factors does not change the fact that Texas suffers significantly ($p<0.01$) greater adverse impacts from floods than does Florida. While Florida also has more porous and drainable soils than coastal Texas, which could potentially reduce ponding during heavy rainfall, controlling for this variable across both states has no meaningful statistical effect. The key variable, then, is the regulatory mandate and resolve to plan for and reduce potential flooding. Florida is simply more prepared to mitigate against flood risks and we show through empirical analysis that mitigation works.

Does this mean that the probability of losing one's home or life in a flood is more a function of geography than anything else? Yes, because the motivation to reduce

Table 8.1 *Summarizing effects of key variables on flooding and flood damage*

Variable	Measure	Direction	Flooding	Flood Damage
Natural environment				
Precipitation	Average surface precipitation recorded by NESDIS weather stations on the *day of* the flood event	+	***	***
Precipitation day before storm	Average surface precipitation recorded by NESDIS weather stations on the *day before* the flood event	+	—	***
Duration of storm	The length of a flood event in days	+	***	***
Floodplain area	Area (in square meters) of 100-year floodplain	+	—	*/*
Stream length/density	Total length of all orders of streams divided by jurisdictional area	+	Nonsignificant	Nonsignificant
Human and built environment				
Impervious surface	Total square meters of impervious surface based on a statewide land cover GIS layer created from Landsat Thematic Mapper satellite imagery	+	*	*
Population density	Population per square kilometer by unit of analysis	+	Nonsignificant	***
Wetland alteration	Cumulative federal wetland permits issued under Section 404 of the CWA	+	**	***
Socioeconomic				
Income	Median household income for unit of analysis	–/+	—*	+ (nonsignificant)
Housing values	Median home values for unit of analysis	+	—	***
Housing value Density	Aggregate county housing value divided by the total county area less the acreage in conservation uses	+	—	**

Political/administrative

State	FL= 0; TX= 1	+	***	***
Adjacent damage	Total property damage in all adjacent counties where flooding from an event occurred	+	—	***
Structural mitigation	Sum of five possible flood mitigation techniques implemented by a local jurisdiction on a 0–5 scale	—	—	Non-significant
Non-structural mitigation	Sum of 15 possible flood mitigation techniques implemented by a local jurisdiction on a 0–5 scale	—	—	***
Organizational capacity	Sum of 13 possible organizational capacity characteristics present in a local government flood organization	—	—	*

*p<0.1, **p<0.05, ***p<0.01.

the adverse impacts of floods and develop more resilient communities over the long term, in many instances, begins at the state level and is then implemented locally.

It is important to note that the impact of storms will not be confined to a single jurisdiction. Floods are largely regional events that do not adhere to administrative or jurisdictional boundaries. Indeed, in an analysis of flood damage at the county level in Florida, the most powerful statistical predictor was the damage incurred in an adjacent county (Brody *et al.*, 2007a: 330–345). This "adjacency effect" stems from the fact that a single flood event often impacts multiple localities. Decision makers therefore should consider multiple jurisdictions when devising flood mitigation policies and focus on ecological units as opposed to arbitrarily defined counties, cities, or towns. Effective flood planning may require a heavy emphasis on collaboration across parcels, organizations, and jurisdictions within a larger geographic region.

Summary

This chapter demonstrates the influence of several local contextual variables (natural environment, human and built environment, socioeconomic, political/administrative) that are important to consider when constructing an overall model explaining flooding and flood impacts in Texas and Florida. In reality, each variable in our conceptual model cannot be considered alone, but in combination with all others presented in Chapter 5. Only a fully specified, interdisciplinary statistical model can begin to unravel flood problems and provide an adequate explanation for their occurrence at the local level. Table 8.1 shows the direction, significance, and effect of each variable's parceled effect as observed over multiple analyses within the study area. While the influence of each variable is described above, understanding each one's effect in relation to all others is crucial when developing a comprehensive view of the factors leading to flooding and flood damage among local communities.

Part III

What are we learning?

9

Policy learning for local flood mitigation

Most research on flood mitigation assumes it is a policy endeavor that is fixed in time. Longitudinal studies that track how communities adjust their flood policies in response to various stimuli are the exception. However, in reality, local policy making is a long-term process where decision makers constantly revise their strategies to address flooding and other natural hazard concerns. Shifting socioeconomic, political, and geophysical landscapes require that local plans and policies be evolving instruments. The ultimate success of a community in terms of becoming resilient to floods, then, may depend on its ability to learn.

In this chapter, we longitudinally explore the local flood mitigation problem by focusing on how jurisdictions change their policies in response to repetitive flood events. By examining the drivers of policy change, we can better understand the degree to which communities improve their mitigation capabilities in the face of chronic flood events. Furthermore, identifying the levers of learning for flood mitigation will inform communities outside the study area on how best to speed the process of adopting flood reduction policies and take a more proactive approach to mitigating the adverse impacts of floods before they occur.

Initially, we draw upon the adaptive management and policy learning literature as a framework for empirically investigating the topic. Then, we focus on Florida as the learning laboratory in which to track policy change: first, we examine the degree to which communities in Florida alter their flood policies across two generations of comprehensive plans, and, second, we observe how counties across the state change their CRS policies over time in response to multiple drivers including flood damage, socioeconomic factors, and physical risk variables. Identifying levers for flood policy learning can help decision makers understand how to improve their local programs and move toward more resilient patterns of development.

Adaptive management

Adaptive management provides one of the most useful decision frameworks for understanding policy learning for flood mitigation. This approach to decision making recognizes that flood managers must continually respond to changing local conditions and that plans and policies need to be flexible instruments geared toward the unexpected. In this way, adaptive management often treats policies as hypotheses tested through the implementation of specific mitigation techniques (Holling and UNEP, 1978; Lee, 1992: 542; Schön, 1983). In this situation, the policy maker learns from evaluating the successes or failures of management actions. Based on an analysis of this new information, objectives and specific strategies are adjusted accordingly to achieve a more desirable effect. As long as the actions set forth are reversible (which sometimes is not the case), the decision maker turned experimenter can improve his or her program over time (Holling, 1996: 733–735).

An adaptive approach to management recognizes that social ecological systems are constantly in flux and often reorganize themselves once a threshold has been crossed. Building and maintaining a high level of resiliency may help these systems cope with the shock of human-induced disturbances (Walker and Salt, 2006). By following adaptive management guidelines, local planning, emergency management, and environmental agencies can be receptive to the cycles of change in the system (both ecological and human) and respond quickly with appropriate and effective management techniques (Westley, 1995: 391–427). Adaptive management is thus an iterative process of action-based planning, monitoring, and adjusting, with the objective of improving future management actions (Holling, 1995: 3–34; Lee, 1993).

Through an adaptive approach to flood management, local decision makers can learn incrementally and refine policies to reduce the adverse impacts of repetitive floods. For example, development restrictions within a designated buffer along major stream segments to reduce repetitive flood damage could be implemented experimentally. If the policy succeeds in meeting intended outcomes, the hypothesis is affirmed and human safety is protected. If, on the other hand, the policy fails and homes outside the buffer begin to flood, more expensive structural interventions may then be needed. In this way, experiments can bring surprises, but these surprises become opportunities to learn rather than failures to predict (Lee, 1993).

Factors influencing flood policy learning

Ensuring the development of flood-resilient communities clearly requires an adaptive and responsive approach to management. Less clear are the factors contributing to policy learning for flood mitigation at the local level. Identifying the "levers for learning" and subsequent policy changes can help decision makers expedite their own jurisdictional adaptive processes leading to higher quality, more effective flood programs.

Several types of learning have been identified in the literature, depending on the management perspective. For example, May (1998) presented adaptive management as an "instrumental" form of policy learning, where the policy maker uses a technically rational approach to attain stated goals. Instrumental learning takes place when decision makers test the viability of specific policy interventions or, as described above, conduct rigorous policy experiments. According to leading political theorists, the largest influence on organizational learning and policy change is preceding policy (Heclo, 1974; Sacks, 1980: 349–376). The adaptive management process is thus incremental, where the objectives policy makers pursue are based on "policy legacies" or meaningful reactions to previous policies. In other words, the major factor affecting a policy adopted at time 1 is tied to prior policy conditions at time 0 (Hall, 1993: 275–296).

Other factors driving policy learning for flood mitigation are the number and extent of past flooding events (Beem, 2006: 167–182; Sabatier, 1995: 412). Floods are spatially defined repetitive events, allowing decision makers an opportunity to improve policies from one flood to the next. When policies are regularly updated over time, they can reflect the learning that takes place within an organization and community as a whole. Because learning is based on the ability of decision makers to anticipate and adapt to unexpected disturbances within an existing system, damaging flood events may actually expedite the process of policy change (Folke *et al.*, 2005: 441). Natural disasters often focus public attention and induce a reactionary response from decision makers, opening windows of opportunity for policy change (Berke, 2007: 283–295). In this way, floods can be thought of not as simply a disaster, but as a trigger for the adoption of new, more effective policies.

For decades, researchers have shown that flood history is a significant factor in the decision of local communities to adopt new standards, such as those mandated by the NFIP (Luloff and Wilkinson, 1979: 137–152; Moore and Cantrell, 1976: 484–508). More recently, Burby (2003) found that repetitive property losses from floods are a significant predictor of stronger local planning policies. Also, in a study in England and Wales, Johnson *et al.* (2005) showed that the extent of floods acted as a mechanism for local policy enhancements associated with flood mitigation. These studies and others suggest that two types of flood history drive policy learning at the local level: acute and chronic. An acute event could be a hurricane that causes an extensive shock to the social ecological system. In contrast, a chronic event involves repetitive small-scale floods that generate cumulative impacts over time. Both types of flood history can trigger policy change.

Most of the focus on adaptive approaches to management assumes that the experimenter (i.e., flood planner) is a rational individual able to test policy hypotheses and implement the results of the experiment. This is a command and control approach that seeks to optimize the efficiency of the system rooted in traditional natural resource management (Walker and Salt, 2006). However, in the local planning arena (particularly in Florida), the experimenter is almost never a lone technician, but rather a tiny cog in a vast organizational wheel that is further embedded within a larger community composed of a complex social network. Public decision making is usually accomplished through the participation of a diversity of stakeholders, including environmental nongovernmental organizations, neighborhood association, development groups, and individual businesses.

This approach to management is termed "social policy learning," which is based on collaboration among a wide range of stakeholders, rather than on one expert or individual (Heclo, 1974). These "policies with publics" may have greater potential for learning because their adoption involves a constant challenging of assumptions and proposed policies by competing advocacy coalitions (May, 1992: 187–206). Adaptive management may be based on the principles of scientific experimentation, but it is ultimately about collective human values and a political culture that tolerates learning from mistakes.

Based on this premise, it is important to consider the external influences of socioeconomic, demographic, and human capital characteristics on policy learning. Levels of wealth, education status, and population composition may affect the speed of learning or change in flood mitigation efforts as much as does the decision maker in charge. For example, wealthy coastal communities could have a greater interest in ensuring mitigation measures are taken because they contain more expensive homes at risk from flooding. These communities are also more likely to have the financial resources to implement costly strategies such as structural relocation, or drainage improvements. Indeed, the difference between Sanibel Island and Ocala, Florida, in the ability to cope with repetitive floods is starkly different, based partly on large differences in the wealth of the populations living there. These and other characteristics related to social and human capital shape the willingness and capacity of a local jurisdiction to minimize flood risks over time.

One way to observe the learning process described above is to track policy change. Because natural hazard plans and policies at a jurisdictional level undergo revision on a regular basis, it is possible to measure improvements over time. The degree to which localities change their flood programs from one update to the next indicates the extent and speed of collective learning. By matching each learning time period with contextual variables, we can better understand why one jurisdiction may learn faster than another and what the key triggers for programmatic leaps may be. As already mentioned, understanding how to speed the learning process will lead to the development of more sustainable and flood-resilient communities. Despite the importance of adaptive management and policy learning, longitudinal studies that empirically track plans and actions over time are rare. Instead, researchers treat plans and planning problems as isolated incidents occurring in a broader spectrum of public decision making. Single time period or cross-sectional studies are the norm partially because temporal data are difficult to obtain and when they are available, it takes a great deal of effort to measure subtle changes over time.

Policy learning through comprehensive plans

Our initial attempt to address the temporal challenge and measure policy learning associated with flood mitigation examined comprehensive plans in Florida over an eight-year period. We evaluated a random sample of 30 local jurisdictions across the state in 1991 (initially selected for a study by Burby *et al.*, 1997) and again in 1999 against a protocol measuring flood mitigation plan quality. Analyses determined the extent to which the flood mitigation components in the comprehensive plans for each jurisdiction changed over time, and then identified the factors driving communities to adopt stronger flood policies. To accomplish these analytical goals, we re-analyzed data collected for a study conducted by Brody (2003b) to focus on flood hazards.

While previous research provides a conceptual and methodological basis for determining the quality of a plan, few studies have examined how and why plan quality changes over time. Policy learning based on the change in plan content or quality can be measured because plans for the same jurisdiction adapt to new conditions over time. In Florida, this change is driven institutionally by a mandate to revise local plans every seven years. The ability to code and measure indicators within a plan has made it a widely used instrument with which to quantitatively assess the quality of management efforts (see Brody, 2008, for more details).

We conceptualized and measured plan quality by incorporating flood mitigation into existing notions of what constitutes a high-quality plan. As used in past studies of local plans and hazard mitigation, we constructed plan quality as three equally weighted components: a strong factual basis, clearly articulated goals, and appropriately directed polices (Godschalk *et al.*, 1989, 1999). Specifically, the fact base refers to existing local conditions and identifies needs related to community physical development. Goals represent aspirations, problem abatement, and needs that are premised on shared values. Finally, policies are a general guide to decisions (or actions) about the location and type of development to assure that plan goals are achieved (Berke and French, 1994: 237–250).

Together these three plan components comprise a local plan to mitigate the negative effects of floods. Indicators (items) within each plan component further specify the conceptualization of plan quality. The fact base component includes background data on the location and extent of hazard damage including the delineation of flood magnitudes, vulnerable populations, and structural loss estimates. Indicators in the goals plan component cover economic impacts (e.g., reduce property loss and minimize fiscal impacts), physical impacts (e.g., reduce property loss, maintain naturally occurring wetlands), and public interest impacts (e.g., protect human safety, increase public awareness of floods). The policies plan component is the most extensive of the three, including actions associated with increasing awareness, regulations, incentives, reducing structural loss, and recovery from floods.

Each of the 72 indicators was measured on a 0–2 ordinal scale, where 0 is not identified or mentioned, 1 is suggested or identified but not detailed, and 2 is fully detailed or mandatory in the plan. Measures of overall plan quality were calculated by creating indices for each plan component and overall plan quality (Berke *et al.*, 1996: 79–96). Indices were created by dividing the actual score by the total possible score and multiplying by 10 to derive a component scale from 0 to 10. A total flood plan quality score was obtained by summing the three component index scores (creating a scale from 0 to 30). Box 9.1 lists each indicator in our protocol.

Box 9.1 **Flood mitigation plan coding protocol**

Factual base
A. Type of data

1. Delineation of location of hazard
2. Delineation of magnitude of hazard
3. Number of current population exposed
4. Number and total value of different types of public infrastructure (water, sewer, roads, storm water drainage) exposed
5. Number and total value of private structures exposed
6. Number of different types of critical facilities (hospitals, utilities, police, fire) exposed
7. Loss estimations (number and total value) to public structure
8. Loss estimations (number and total value) private structures
9. Emergency shelter demand and capacity data
10. Evacuation Clearance Time Data

Goals A. Economic impacts

1. Any goal to reduce property loss
2. Any goal to minimize fiscal impacts of natural disasters
3. Any goal to distribute hazards management cost equitably

B. Physical impacts

1. Any goal to reduce damage to public property
2. Any goal to reduce hazard impacts that also achieves preservation of natural areas
3. Any goal to reduce hazard impacts that also achieves preservation of open space and recreation areas
4. Any goal to reduce hazard impacts that also achieves maintenance of good water quality

C. Public interest

1. Any goal to protect safety of population
2. Any goal that promotes a hazards awareness program
3. Other (specify)

Actions
A. General policy

1. Discourage development in hazardous areas

B. Awareness

1. Educational awareness
2. Real Estate Hazard Disclosure
3. Disaster warning and response program
4. Posting of signs indicating hazardous areas
5. Participation in flood insurance programs
6. Technical assistance to developers or property owners for mitigation
7. Other (specify)

C. Regulatory

1. Permitted land use
2. Transfer of development rights
3. Cluster development
4. Setbacks
5. Site plan review
6. Special study/impact assessment for development in hazard areas
7. Building standards
8. Land and property acquisition (eminent domain)
9. Impact fees
10. Retrofitting of private structures
11. Other (specify)

D. Incentives

1. Retrofitting of private structures
2. Land and property acquisition
3. Tax abatement for using mitigation
4. Density bonus

5. Low-interest loans
6. Other (specify)

E. Control of hazards

1. Storm water management/watershed treatment
2. Maintenance of structures
3. Other (specify)

F. Public facilities and infrastructure

1. Capital improvements
2. Retrofitting public structure
3. Critical facilities
4. Other (specify)

G. Recovery

1. Land use change
2. Building design change
3. Moratorium
4. Recovery organization
5. Private acquisition
6. Financing recovery
7. Other

H. Emergency preparedness

1. Evacuation
2. Sheltering
3. Require emergency plans

4. Other (specify)

From Brody (2003b) "Are we learning to make better plans? A longitudinal analysis of plan quality associated with natural hazards," *Journal of Planning Education and Research*, **23** (2), 191–201, with permission.

Findings

The total flood mitigation plan quality for our sample of jurisdictions in Florida increased significantly (based on paired t-tests) from 1991 to 1999. While a 25% increase over eight years is statistically significant, in 1999 the average total score was only 5 on a scale of 0–30, indicating the need for additional learning. The improvements in flood mitigation plan quality were driven almost entirely by the policies plan component, which increased by over 90%. The fact base and goals components, by comparison, showed only modest, nonsignificant gains despite the fact that we were evaluating what were supposed to be major plan revisions.

Our findings indicate that the fact base of a plan is the most difficult component to update. Major revisions require additional studies, monitoring of existing environmental conditions, GIS map preparation, and data gathering based on long-term monitoring programs. Fact base indicators seem to take longer to "catch up" to the other plan components due to the necessary commitment of time and financial resources to keep these indicators current. A slower learning curve for a factual base should not be overlooked because this component acts as the foundation of a plan, driving goals and policies to mitigate floods. Without supporting data and analysis, a plan may fail when it comes to implementation and overall effectiveness.

In contrast, the actions component of our flood mitigation protocol made the most improvement over the eight-year study period, which may be the strongest indication that policy learning and adaptive management is taking place in Florida. Localities strengthened their ability to both mitigate and recover from flood hazards. Specifically, comprehensive plans improved policies associated with discouraging development in hazardous areas as well as participating in federal flood insurance programs. In particular, the FEMA CRS (examined in the next section) became more widely utilized as a vehicle for mostly non-structural forms of mitigation. Local jurisdictions also embraced what are considered more innovative land use policies, such as transfer of development rights, cluster development, and impact fees. The addition of these policies reflects the growing acceptance of planning at the local level in Florida and the advancements possible with a second generation of adopted plans and policies. Hurricane Andrew, which made landfall in south Florida in 1992, combined with increasing pressure from the FEMA, also played a role in improving preparation for future flood-related disasters.

Drivers of change

We expected to notice improvements in the sample of comprehensive plans as they went through an updating process, partially because the starting point scores were so low. More important are the reasons for these changes and understanding why some localities improve more than others. To address this issue, we constructed and analyzed an explanatory model using OLS regression analysis to predict the variation in policy change from 1991 to 1999. Contextual data were selected based on the existing literature on policy learning and plan quality, then collected from various secondary sources as well as interviews with planning directors and staff at each location (Brody, 2003b: 191–201). In the end, we included the following variables in our model:

- plan quality at the initial time period;
- population growth;

- the number of citizen groups participating in the planning process (citizen participation);
- the change in demand for development in hazard-prone areas;
- reported repetitive property losses in 1990 (chronic losses);
- change in the number of planning staff devoted to hazard mitigation (planning capacity);
- change in commitment of elected officials to mitigate natural hazards (commitment).

As anticipated, the biggest predictor of plan quality in 1999 was plan quality in 1991. Jurisdictions clearly build on previous policies to establish what have been called "policy legacies" (Weir and Skocpol, 1985: 107–168). These legacies become institutionalized within planning organizations and tend to carry over to future plan updates. However, if a locality adopts a high-quality plan from the outset, there may be less room for drastic improvements in the future.

Increasing amounts of repetitive damage to specific properties from floods was also a statistically significant predictor of hazard plan quality in 1999. Because these chronic losses are linked to specific locations, they tend to generate interest in a policy response. For example, Sarasota County, located on the west coast of Florida, responded to repetitive flood events by purchasing and protecting several of the most vulnerable parcels within its jurisdiction. A responsive and committed Sarasota planning office led to the establishment of a very strong flood mitigation program and one of the best CRS ratings in the state. In contrast, general issues addressed during the development of a comprehensive plan, such as sustainable development, are more difficult for decision makers to personalize with meaningful actions.

Increasing demand for development in hazard-prone areas, such as the 100-year floodplain, was another factor contributing to flood policy learning via comprehensive plans from 1991 to 1999. In this instance, mounting pressure to develop significantly *reduced* the resulting quality of local plans. This study found that political and economic pressures to develop in profitable, but vulnerable, areas can overwhelm the public desire to protect critical natural resources, personal property, and at times even human life. The pressure to allow development on prime coastal real estate in Florida for residential and tourism uses is so great that sound planning for flood resilience is sometimes cast aside. High-density urban development on beachfronts of Fort Lauderdale, Clearwater, and other coastal cities demonstrates the overwhelming financial will to develop vulnerable areas without consideration of the natural environment or public safety (see Deyle *et al.*, 2008: 349–370 for a more recent analysis).

Several other variables included in the model were not significant predictors despite theoretical and empirical evidence to the contrary. One would expect that high planning capacity and strong political commitment to mitigate natural hazards would contribute to an improvement in plan quality over the study period. Such "non-findings" raise the question of how much time must pass before these nonsignificant factors contribute to policy learning. If we examined plans over 10, 15, or 20 years, would that be enough time for political commitment to filter down to the planning staff level? Would it be enough time for an increase in planning staff to improve the quality of adopted plans? These questions suggest that there might be different learning time thresholds for each factor that plays a role in policy learning. To better understand the policy learning process, longer time periods must be studied.

Assessing the FEMA's CRS: a program in motion

Florida's comprehensive plans are meaningful instruments with which to examine policy learning because they form the regulatory basis of local flood mitigation techniques. However, several issues make it difficult to conduct a thorough longitudinal analysis. First, these plans are only updated (aside from minor variances) every seven years, making it difficult to monitor multiple time periods. In fact, most jurisdictions are using only a third-generation plan, providing only a few points to assess policy change. Second, long time lapses between plan updates increases the possibility of "history threats" or various occurrences that may confound our ability to explain why learning takes place. Third, while comprehensive plans are legally binding policy instruments in Florida (see Chapter 4), it is uncertain whether they are fully implemented. In fact, previous studies have found local development patterns frequently violate the original intent of adopted plans (Brody and Highfield, 2005: 159–175; Deyle *et al.*, 2008: 349–370).

To dig deeper into the topic of policy learning for flood mitigation, we turn our attention once again to the FEMA CRS. In a separate analysis, we examined the change in local flood mitigation policies in Florida from 1999 to 2005 using CRS activities as benchmarks for learning. We followed all 18 mitigation activities and their four series (see Chapter 4 for more details on the program) on a yearly time-step for every local jurisdiction in Florida participating in the FEMA program (this case study is based on analysis by Brody *et al.*, 2009c: 912–929). This analytical approach provides a larger sample of jurisdictions and many more time periods then the case study above, enabling us to better understand the degree to which policy change occurs and the factors driving this learning process.

The CRS provides an ideal policy mechanism with which to monitor and explain flood mitigation efforts over time for several reasons. First, because each activity under the program must be implemented, CRS certification demands a level of commitment from participating jurisdictions that may not be present in comprehensive plans. Every participating community is evaluated by external reviewers to make sure mitigation activities are implemented. Second, CRS localities must recertify by October 1 of each year that they are continuing to put into practice the activities for which they have received credit points. Third, a CRS community can adjust its application each year by adding or enhancing mitigation activities so that it can earn more credit points and move to a higher class rating (to obtain a greater discount on flood insurance premiums). In this way, the CRS is an adaptive flood mitigation program, enabling a community to change its policies over time in concert with changes in the geophysical, political, and socioeconomic landscapes.

Florida provides an ideal focal state within which to test the policy-learning process due to its high rate of participation in the program (52 out of 67 counties at the time of the analysis) coupled with the persistent risk of damaging flood events. With such a rich set of data, we can begin to address two important questions associated with policy learning: (1) How much are jurisdictions changing their activities on a yearly basis; and (2) What are the factors facilitating this change?

To address the first question, we measured the change in CRS scores by series over the seven-year study period. While most counties in Florida have one CRS community, in many cases, a jurisdiction within a county (e.g., city or town) earns separate scores for their own mitigation activities. In these instances, we population-weighted and summarized the mitigation activities of a nested municipality and its county. Using this procedure, our adjusted county CRS scores reflected the number of people that directly benefit from specific flood mitigation efforts. Once we derived a score for each CRS jurisdiction in Florida, we calculated its depth as the total points earned divided by the maximum points available (for a more detailed explanation of our measurement logic, see Brody *et al.*, 2009c: 912–929).

Figure 9.1 illustrates CRS depth scores by series and year, indicating the degree to which policy learning or change occurs for specific sets of policies. From 1999 to 2005, there was an upward trend in flood mitigation activity across all CRS series. Overall, communities accumulated the most points for public information activities (series 300), earning (on average) over 28% of the total points available. Over the time period examined, localities made the greatest improvement under series 400 (maps and regulations) activities. These scores increased from 5.38% in 1999 to 11.00% of available points in 2005. In contrast, participating jurisdictions scored substantially lower for series 500 (damage reduction) and 600 (flood preparedness) activities. For example, series 600 scores only moved from 6% in 1999 to 7% of earnable points in 2005.

While CRS scores trend upward over time, the starting point for most of the jurisdictions was very low, leaving vast room for improvement. For example, when combining points for all the series, Florida communities earned less than 10% (on average) of the total CRS points that could be obtained. The highest scoring

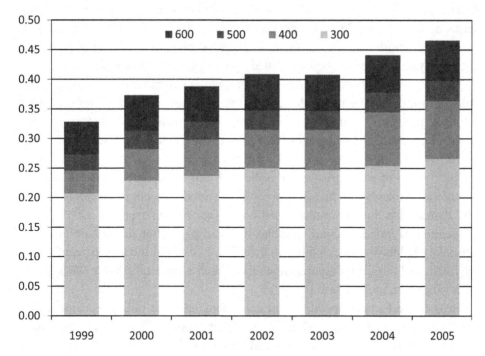

Figure 9.1 CRS class scores on a yearly basis.

counties receiving the largest insurance premium discounts are Charlotte, Lee, St. Johns, Manatee, and Hillsborough, which earned between 14% and 17.5% of total CRS points. These top performers appear clustered geographically on the west coast of Florida, around the Charlotte Harbor Estuary and the Tampa Bay region.

Based on the descriptive results above, it seems that localities in the CRS program are pursuing a form of least-cost learning in that they disproportionately focus on point-earning activities that are less expensive, more politically viable, and generally easier to obtain. In particular, CRS participants appear to favor series 300 and 400 activities, which involve mostly information dissemination, public outreach, and strengthening of existing regulations. In contrast, activities in the 500 and 600 series that may require relocation of structures or drainage investments requiring greater fiscal commitments are adopted less frequently. The high capital cost of structurally related activities most likely explains why localities in Florida are more inclined to implement non-structural solutions.

While it is human nature to take the least-cost approach to achieving a desired end (in this case cheaper insurance premiums), the pursuit of "low-hanging fruit" may reduce the effectiveness of flood mitigation programs. For example, in 2005, almost 74% of all points earned were based on series 300 and 400 activities, which is 28% higher than the proportional weight assigned to these activities per program specification. Scores, on the other hand, deviate below the proportional weight of series 500 scores by 23.49% (see Brody *et al.*, 2009c: 912–929). Adjusting the weights of certain activities to increase potential rewards may help to better balance the type of activities being pursued by participating communities.

To address the second research question (why do communities learn?) we constructed a quantitative model to explain the variation in CRS scores over time. Multiple predictor variables were identified and measured based on existing conceptions of factors influencing learning as described above (see Brody *et al.*, 2009c: 912–929 for more details). Under the category of *hydrology*, we measured the percentage of each jurisdiction in the 100-year floodplain and the total length of streams in meters. For *flood history*, we measured frequency and extent of property damage for flood disasters based on 10-year rolling averages. Lastly, *socioeconomic and human capital* variables included population density, dollars saved per policy holder by participating in the CRS, household income, percent college educated, and per capita non-profit assets in each community. Table 9.1 describes each variable in more detail.

To isolate effects of specific local conditions on flood policy learning via the CRS within a locality, we used panel regression models (52 panels, 7 years) on an annual time-step. Models were analyzed for each CRS series as well as for overall scores. This analytical approach enabled us to identify the levers for learning for flood mitigation in Florida.

While there were some variations in the results across CRS series, consistent patterns emerged from the data in terms of specific factors influencing policy change over time. First, both hydrological variables generally decreased CRS scores. For example, increasing percentages of 100-year floodplains significantly decreased scores from 2 to 9%, depending on the series. In fact, jurisdictions with at least 25% of land area in the 100-year floodplain performed significantly worse across all activity series (Brody *et al.*, 2009c: 912–929).

There are several plausible explanations for the negative effects of the hydrological variables, specifically within the floodplain. First, the floodplain could act as a deterrent for development because communities want to reduce their risk of exposure, so there is less of a need to adopt extensive flood mitigation policies. Second, mitigating the adverse impacts of floods where there are large areas of floodplain requires more expensive and politically less desirable policy interventions. In these cases, the financial benefits of lower insurance premiums may be outweighed by the high cost of obtaining more CRS points. Localities under these conditions may be more likely to stall their policy efforts. Given the fact that all of our measures of wealth are negatively correlated with floodplain percentage (p<0.001), even if communities with large amounts of floodplains wanted to increase their point totals, they typically do not have the financial resources to

Table 9.1 *Policy learning predictor variable measurement*

Variable	Variable measure
Hydrology variables	
Floodplain percentage	Total land area of a county in the floodplain divided by the total land area (in square kilometers).
Stream length	Total length of streams in a county area (in meters).
Flood history variables	
Flood frequency	Ten-year rolling average of the total annual number of flood disasters recorded in a county.
Flood property damage	Ten-year rolling average of the total annual flood-caused property damage recorded in a county in $10 000 increments (in year 2000 inflation adjusted dollars)
Socioeconomic variables	
Population density	Total population divided by country area (in square kilometers). Values for 1990 and 2000 Censuses are used to estimate intervening years, assuming equal interval of change
Reduction per policy holder	Total dollars saved divided by the total number of FEMA National Flood Insurance Program policy holders
Human capital variables	
Nonprofit assets per capita	The total assets reported by all number non-profit organizations of tax-exempt status with $25 000 dollars in gross receipts required to file Form 990 with the IRS in a county divided by the total population
Median household income	The sum of money received in a year by all household members 15 years old and over. Values for 1990 and 2000 censuses are used to estimate intervening years
Percent college educated	Number of persons age 25 and over with a bachelor's, master's, professional, or doctorate degree divided by the total population 25+ years of age. Values for the 1990 and 2000 Censuses are used to estimate intervening years

IRS, Internal Revenue Service.
Adapted from Brody *et al.* (2009c) "Policy learning for flood mitigation: a longitudinal assessment of CRS activities in Florida," *Risk Analysis*, **29** (6), 912–929, with permission.

implement the necessary policies, which could involve land acquisition, relocation, drainage system maintenance, etc.

Money saved per policy holder is the most consistent significant factor among socioeconomic variables. This finding may be due to local decision makers responding to the per capita financial gains that accrue to their constituency from engaging in mitigation activities. Increasing population density also contributes to significantly higher CRS scores. As population becomes more concentrated within a jurisdiction, there is greater justification for pursuing extensive mitigation techniques to protect property and human life.

Finally, wealthier and more educated communities appear to learn faster. Localities with a large amount of financial resources and expertise can invest in more extensive flood mitigation programs, which results in higher CRS scores over time. Our socioeconomic and human capital findings give support to the idea that local jurisdictions are very aware of potential economic gains to be made by participating in the CRS. Localities in Florida have adopted flood mitigation strategies as the expected benefits from doing so increase by way of monies saved per policy holder. The larger the potential savings, the more incentives localities have to pursue flood mitigation techniques.

Flood history also plays a role in triggering policy learning at the local level. Both the frequency and extent of flood events are positively significant for most series, although the size of the effects is fairly modest. This result tells us that experience with the negative impacts of flood hazards is important to facilitate policy change over time. It seems the number of flood repetitions is more influential in stimulating policy makers than the severity of property damage caused by storms. What our research could not answer is the question of how many times a community has to be hit by a flooding event, or how large an economic impact floods must have, before flood mitigation policies are implemented. Because this is likely a key factor in policy learning, we encourage future research to investigate this issue in more detail.

Summary

This chapter has shown that plans and policies to reduce the adverse impacts of floods involve an iterative learning process. We cannot view mitigation decisions made by communities as fixed in time; rather, they are constantly in motion as their leaders and residents struggle to adapt to ever-changing conditions. The key, then, for reducing exposure to floods is to better understand how to expedite the policy learning process and how to facilitate the development of communities that are resilient to the surprises that nature has in store. Our analysis is simply a first step in unraveling the nature of policy learning for floods and we hope this chapter leads to more empirical research on the topic involving larger samples and longer time periods.

10

Local case studies in Texas and Florida

Up to this point our study has been conducted largely at a "high altitude," drawing conclusions across multiple jurisdictions at broad spatial scales. Studies of this kind are important because they allow us to make generalizations (in this case at state levels) that can inform a larger body of decision makers. So far, we have also offered quantitative empirical evidence on the causes and consequences of flooding among localities in the U.S., which is often absent in the scholarly and technical literature. What we miss, however, is a fine-grained, qualitative assessment of activity within specific communities. Unraveling the intricate socioeconomic, political, and environmental characteristics influencing a locality's decisions related to flood mitigation can lead to explanations that more comprehensive studies may miss.

To supplement our findings based on large-sample analyses and statistical models, we select five local jurisdictions: Galveston County, Texas; Freeport, Texas; Manatee County, Florida; St. Petersburg, Florida; and Palm Beach Gardens, Florida (see Figure 10.1), and provide a more detailed, integrated description of what is taking place in terms of coping with chronic floods. We selected these localities based on their variation in flood losses and mitigation policies, as well as on diversity in local conditions. These "profiles" include physical, socioeconomic, and policy-based contextual characteristics that exemplify the themes presented in previous chapters. The cases also provide an opportunity to explore how multiple factors come together to form an overall picture of flood impacts and mitigation responses. Each profile is written based on information taken from surveys, flood loss databases, the U.S. census, planning documents, and conversations with local decision makers.

Galveston County, Texas

Galveston County is situated on the upper Gulf Coast of Texas, just southeast of the sprawling metropolis of Houston. This jurisdiction encompasses Galveston Island and Bolivar Peninsula, the City of Galveston, Texas City, and Clear Lake, among

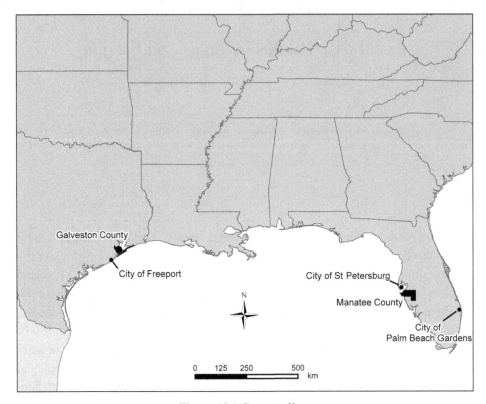

Figure 10.1 Case studies.

other notable towns. It contains a diverse economic base, including oil and gas refineries, NASA mission control, recreational beaches, and several tourist destinations.

Throughout this book, we have illustrated conditions under which localities are vulnerable to the adverse impacts of floods and Galveston County seems to possess them all. First, over 46% of its land area is located within the 100-year floodplain. Second, the clay-dominated soils in this region make natural infiltration of rainfall into the ground problematic. Third, in addition to the lack of permeability in the underlying soil structure, the topography in the region is extremely flat and the elevation is barely above sea level (most of the area is below 8 ft (2.4 m)). Finally, the region's natural hydrology is dominated by wetlands, bayous, and other natural features meant to collect, store, and slowly disseminate water. These geophysical conditions make Galveston County extremely sensitive to disturbances on the landscape, such as human development. Even minor disruptions in storm-water runoff can lead to consequences associated with flooding.

Recent development in Galveston County can be best characterized as suburban and ex-urban residential sprawl. Large subdivisions that cater to residents commuting north into Houston for work are being built adjacent to Interstate 45. As of the 2000 U.S. Census, the population for Galveston County was just over 250 000, representing a 15% increase from the previous decade. Although 2010 Census figures are not yet available, 2009 population is estimated at 287 000 and future increases are expected to be steep. Approximately 500 000 additional persons are projected to move into this area in the next couple of decades.

With people come rooftops, roads, retail developments, and other impermeable surfaces that act as a major trigger for flooding (see Chapter 5). Using satellite imagery, we calculated that nearly 20% of Galveston County comprised impervious surfaces, as of the year 2000. Pavement associated with sprawling development patterns increases storm-water runoff and reduces the ability of hydrological systems to naturally store rainfall events. Impervious surfaces surely played a major role in explaining over $18 million of insured flood losses claimed from 1996 to 2007 within Galveston County alone. According to FEMA records, over $6.4 million was paid out for repetitive flood damages, where property owners situated in the most vulnerable areas file multiple claims that may cumulatively exceed the market value of the building itself.

As discussed in Chapter 6, when the addition of impervious surfaces corresponds to a loss in naturally occurring wetlands, the risk of flooding and flood damages increases. From 1991 to 2003, we calculated that 1185 wetland alteration permits were issued in Galveston County under Section 404 of the CWA. We previously isolated the effect of wetland permits on flood damage using statistical controls, but a glance at the record for Galveston County illustrates the general trend. For example, in April 1997, a rainfall event of just 0.09 inches (0.228 cm) caused $5000 in property damages. At this time, the USACE had cumulatively issued 546 wetland permits. In September, 2000, the same amount of precipitation caused $100 000 in damages during which the number of permits on record had increased to 921.

Galveston County's policy and planning response to chronic flooding has focused primarily on structural interventions. Responses to the flood mitigation survey indicated extensive use of building retention/detention ponds, levees, clearing of debris, and hardened channels. In comparison, non-structural techniques are employed much less frequently. The lack of attention to non-structural mitigation is emphasized by the fact that Galveston County is not a participant in FEMA's CRS (although several municipal jurisdictions within the county are), which emphasizes "softer" solutions to reducing the adverse impacts of floods. The county does, however, participate in the NFIP under which it passed the required floodplain management plan in 2002.

The plan's provisions meet the minimum standard for NFIP participation (e.g., new construction must have its lowest level above the floodplain or 18 inches (45.7 cm) above ground), but the regulations do not exceed those required by the NFIP (see Galveston County Floodplain Management Plan, 2002). In many ways this is a missed opportunity to engage in more effective mitigation actions, such as requiring residential freeboard (a factor of safety usually expressed in feet above a flood level for purposes of floodplain management) that may further reduce future flood damages. Galveston County is also part of the Houston–Galveston Area Council Hazard Mitigation Plan. Under the Disaster Mitigation Act of 2000, states and local jurisdictions are required to develop hazard (including flooding) mitigation plans as a condition of eligibility for hazard mitigation grants from FEMA. A recent analysis of this plan assigned an overall quality score of 44 out of a possible 100. More specifically, there were four structural mitigation and two natural mitigation actions adopted within the plan (Peacock *et al.*, 2009).

Overall, Galveston County's pursuit of structural flood mitigation strategies may be appropriate given the number of people and structures already located in vulnerable areas. However, as parts of the county experience rapid development, such as in League City, complementary non-structural approaches that include buffers, land acquisition, and directed development would help the area become safer, more resilient to flood hazards. A general reluctance to engage in land use planning to reduce flood impacts, and lack of enabling legislation at the county level, could magnify losses as more people and structures are exposed to risk, an unavoidable occurrence given the underlying physical conditions.

Freeport, Texas

Freeport is a small town located just inland from the Gulf of Mexico in Brazoria County south of Houston. As of the 2000 census, the population was approximately 12 700, which represented an 11.5% increase over the previous decade. Since 2000, the population has actually decreased somewhat (2009 U.S. Census estimates place the population at 12 618), although future growth is expected to add to its 4800 housing units. Freeport's economy is driven by industrial production facilities, along with a strong recreational fishing base. Most notably, Dow Chemical owns and operates a major plant on the harbor side of the town, which is ringed by a levee system to protect against flooding. Thus, while the population is small, the town's industrial production importance is far reaching. As a community, the median household value and income is below the state average. Unemployment and those living below the poverty line are above the state and national averages.

What makes Freeport interesting from a flood perspective is that it is vulnerable to both rainfall and wave-based events. Almost half of the town's area falls within the 100-year floodplain, making future development problematic. Freeport is also vulnerable to storm-surge, coastal flooding, and sea level rise, given its proximity to the Gulf. In fact, most of the jurisdiction is at or below 1 ft (0.30 m) above sea level. Due to its high degree of vulnerability, Freeport has experienced large amounts of flood damage for a community of its size. From 1996 to 2007, the town claimed over $3.5 million in insured losses, the bulk of which occurred in 2002 from several severe storm events (including Tropical Storm Fay).

Aside from the levee surrounding the chemical plant, Freeport's response to flood threats focuses mostly on non-structural approaches to management. For example, survey responses cite extensive use of zoning, setbacks, land acquisition, and training of local residents to reduce the impacts of flood events. Freeport implements these mitigation techniques through a Flood Damage Prevention Ordinance (Chapter 153 of the local code), which seeks, among other methods, to "control filling, grading, dredging, and other development which may increase flood damage" and "prevent or replace the construction of flood barriers which will unnaturally divert flood water or increase flood hazards of other lands." Specific standards contained in the ordinance include a provision requiring all new residential construction in the 100-year floodplain to be located at or above base flood elevation, again the NFIP minimum standard, as it also prohibits the alteration of sand dunes and mangrove stands, which decreases the magnitude of flooding and related coastal damage.

Construction practices and building codes in Freeport are also among the best in coastal Texas. The jurisdiction adopted the 2003 IRC/IBC building code with required inspections in an attempt to ensure structures can withstand not only flooding events based on both precipitation and wave action, but also high winds associated with tropical events. Recently, Freeport was acknowledged by FEMA for its best building practices associated with a Habitat for Humanity initiative. Since 1991 this organization has built 66 new homes that exceed the existing local codes. Not only do these builders use metal clips, precise nail patterns, and extra bracing structures to fortify homes, they also avoid the floodplain altogether. In fact, the minimum Habitat for Humanity requirement for building elevation is 1 ft (0.30 m) above the 500-year floodplain (FEMA, 2008), a policy many communities would benefit from in the long term.

Local examples, such as the Habitat for Humanity program, reflect the strong commitment of residents and public officials to developing a flood-resilient community. According to local building officials, there is high interest from elected officials that filters down to the household level to mitigate flood impacts. Sharing information and resources is another strong attribute among Freeport government officials that helps build resiliency. Finally, financial commitment is a key aspect of building organizational capacity and implementing flood mitigation techniques (Brody *et al.*, 2009b: 167–184). According to survey responses, Freeport has dedicated two full-time professional staff members to flood mitigation activities, which is a significant number considering the size of the jurisdiction.

Freeport's emphasis on non-structural flood mitigation strategies has helped it become a viable community despite its apparent vulnerability to floods. A strong local code combined with a culture of commitment and collaboration are key aspects in protecting both property and lives. Interestingly, Freeport is not a participant in the FEMA CRS, even though it seems its activities would meet the program's basic criteria. Receiving a CRS rating would not only recognize the town for its mitigation efforts, but also enable it to pass on an insurance premium discount to residents living in the floodplain.

Manatee County, Florida

Manatee County is situated on the west coast of Florida, just south of Tampa and St. Petersburg. The jurisdiction experienced population growth in the 1990s of approximately 25%, and as of 2008, there were over 264 000 residents (over 70% of whom lived in unincorporated areas). In 2008, there were approximately 83 000 owner-occupied houses or condos with a median value of $228 200, which is slightly above the state average. Manatee County experienced a boom in residential construction from 2000 to 2005, during which over 24 000 building permits were issued for single-family new home construction alone. Around the same time period, the USACE issued about 150 wetland alteration permits to partially accommodate this growth. Given Manatee's beaches and various natural amenities, the area is a prime spot for retirees and second-home owners.

Manatee's coastal location and lack of topography make it vulnerable to flooding, particularly from tropical storms and hurricanes coming from the Gulf of Mexico. While less than 19% of the county's area is within the mapped 100-year floodplain, sudden deluges of precipitation during the wet season create flooding problems. Tropical storms Gabrielle (2001), Bonnie (2004), and hurricanes Charley, Frances, Ivan and Jeanne (all in 2004) took an added toll in terms of flood-related losses. According to the SHELDUS, from 1996 to 2007, Manatee County suffered almost $13 million in overall damage from flooding events. Given the populace of the jurisdiction, this figure comes out to only about $50 per person (compared to $1156 for Galveston County, Texas). During the same time period, FEMA paid out approximately $7.3 million in flood insurance claims, a relatively low figure given the county's size and population density.

Manatee County has managed to keep its losses comparatively low in the face of numerous severe storms, perhaps because of its strong commitment to flood mitigation. For example, it had the fifth-highest mitigation score for the entire Texas–Florida flood mitigation survey, which received responses from 173 jurisdictions.

Manatee distinguished itself among surveyed localities through its extensive use of non-structural techniques, such as zoning, protected areas, setbacks, and educational programs. An indicator of its dedication to flood mitigation is its participation and success in FEMA's CRS. Manatee entered the program in October of 1991 and since then has risen to a class 6, which translates into a 20% insurance premium discount for residents living in the floodplain and a 10% discount for those located outside the floodplain. In 2005, the county's 18781 policy holders saved over $1.5 million that year alone.

Examining Manatee County's CRS activities more closely further reveals its high level of engagement in terms of protecting its citizens from the adverse impacts of floods. Its overall point total of 2238 is the eighth highest in Florida and in the top 5% in the country. Manatee scored particularly high in addressing storm-water issues. In terms of drainage system maintenance, such as clearing debris from channels, it scored 294 out of a possible 330 points (the national average score is 232). The jurisdiction also performed extremely well under the category of higher regulatory standards, under which new development is provided more protection than that of the NFIP's minimum requirements. For example, 12% of Manatee's higher regulatory standards points are the result of protecting floodplain storage capacity through regulations that require new developments to provide compensatory floodwater storage. Another 19% of this category's point total is the result of adopting and enforcing strong building codes. Overall, Manatee accumulated 581 points for this activity for which the national average is only 166. Finally, information dissemination, assistance, and public outreach were other strong components of the locality's flood mitigation program. For example, it was awarded 66 out of a possible 71 points for providing technical assistance to interested property owners and contractors. For outreach projects that notified residents of flood hazards, the availability of flood insurance, and/or flood protection methods, Manatee scored 167 out of a possible 380 points, where the national average is only 90.

Manatee County's flood preparedness and mitigation policies are activated through its comprehensive plan, the cornerstone of land use decisions at the local level in Florida. Comprehensive planning is where the so-called "rubber hits the road" when it comes to flood mitigation activities, and unless other plans and programs are folded into this central document they may not be implemented. Manatee County's plan was first adopted in 1989 and has since undergone several iterations. The current version seeks an avoidance strategy for flood impact reduction by guiding development away from sensitive and vulnerable areas. For example, a main objective of the plan strives to "limit development type, density and intensity within the Coastal Planning Area and direct population and development to areas outside of the Coastal High Hazard Area to mitigate the potential negative impacts of natural hazards in this area" (Objective 4.3.1). One policy under this objective requires clustering of development to "protect coastal resources from the impacts of dock accesses, runoff from impervious surface and to minimize infrastructure subject to potential storm damage" (Policy 4.3.1.5).

Other policies maintain construction setbacks, minimize disturbance of natural shorelines, and direct public infrastructure outside of flood-prone areas. Manatee is also one of the few jurisdictions in our study that recognizes the value of naturally occurring wetlands for flood attenuation. Specifically, its plan seeks to protect wetland systems to "maintain control of flooding and erosion through storage of agricultural and urban runoff in wetland areas" (Policy 3.3.1). Based on our analysis of satellite imagery, over 35% of the county's area comprises wetlands.

Manatee County is a prime example of an integrated non-structural approach to reducing the adverse impacts of flooding. Utilizing Florida's comprehensive planning mandate as a vehicle, the county has adopted and implemented a range of land use and growth management tools to keep people and property out of harm's way. In fact, Manatee scored eighth highest in an evaluation of 53 local plans in Florida for their strength in mitigating floods (Kang, 2009). Its mitigation initiatives are also reflected in a very high CRS point total and corresponding class, earning residents millions of dollars in insurance premium reductions. All of these programs and policies have contributed to a low rate of per capita flood damage compared with other coastal jurisdictions and a more resilient community in which to reside.

St. Petersburg, Florida

St. Petersburg, located on the western-central coast in Pinellas County, is the fourth largest city in Florida with approximately 245 300 people (as of 2008). The city's close proximity to Tampa Bay and abundant sunshine has helped it become a prominent destination for tourists and second-home owners. Population growth since 1990 has been only about 3%, since much of the development in this urban area occurred in the 1970s. Based on 2000 U.S. Census estimates, there were 124 618 housing units with a median value of $81 000. The largest number of building permits issued for single-family home construction over the last decade was in 2005 with 925. Since that time, new development within the city has either flat-lined or declined.

St. Petersburg's geographic features make it a very vulnerable location for development, which has undoubtedly contributed to its mounting flood damages. Over 45% of its land area (almost 16 000 acres [6475 hectares]) falls within the 100-year floodplain, much of which has been converted to impervious surfaces. In fact, the city is now over 94% built out, one of the highest levels in Florida. Seventy-three percent of floodplain areas are currently developed and 23% are held under preservation status. All of the 26 major drainage areas outfall to tidal water either directly through overland flow or a network of drainage-ways and storm sewers.

As we have shown in previous chapters, alteration of floodplain areas with paved surfaces reduces the ability of rainfall to infiltrate the soil and increases runoff (Schuster *et al.*, 2005: 263–275). For example, one study found that as the percentage of impervious surface within a drainage basin increases to 10–20%, corresponding runoff doubles (Arnold and Gibbons, 1996: 243–258). As we have previously demonstrated, development in flood-prone areas often leads to property losses. Indeed, St. Petersburg incurred almost $30 million in insured flood damage from 1996 to 2007. The bulk of this damage occurred in 1996, during which Tropical Storm Josephine overwhelmed storm-water systems and inundated residential areas. During this time period, St. Petersburg earned the distinction of Florida's top locality for repetitive flood losses. In general, the per-capita flood damage rates for this city are higher than other places analyzed in our study.

St. Petersburg's high rate of property damage from floods is most likely a result of several factors, primarily its high floodplain resident density and below-average flood mitigation efforts. The city had the lowest overall mitigation score in our survey across two states, mostly for its lack of attention to non-structural measures (which one could argue are more difficult to adopt due to the fact that the city is practically built-out). St. Petersburg has been a participant in FEMA's CRS since 1992 and worked its way up to a class 7 during the time this case was written. The jurisdiction scored particularly high for its public outreach activities (activity 330), which involve sending information to residents about flood risks in their neighborhoods. A prominent section of the City's website is devoted to disseminating information on flood risks and responses. It also scored above average for storm-water management regulations (activity 450) and drainage system maintenance (activity 540). Finally, St. Petersburg was awarded 135 points (national average is 93) for its flood warning program (activity 610) that entails providing early flood warnings to the public and having a detailed flood response plan keyed to flood crest predictions. Considering the extent to which the city is developed, the community is likely doing as much as feasible from a CRS and general mitigation standpoint.

As shown in Chapter 7, integration of flood mitigation techniques into local comprehensive plans is a gateway for reduced property damages from flooding events. In a flood plan quality evaluation in Florida (Kang, 2009), St. Petersburg scored slightly above average, but lower than 27 other coastal jurisdictions (most with lower populations and floodplain area). The plan (adopted in 1996) scored lower than the average of the 53 other jurisdictions for using land use and zoning tools to mitigate floods, such as wetland regulations and down-zoning of floodplains. In contrast, St. Petersburg's comprehensive plan overachieved in terms of clearing debris and storm-water management (these findings are consistent with performance under the CRS).

A history of development in flood-prone areas and lack of attention to mitigation programs has left St. Petersburg with a disturbing record of damages caused by floods; our research touches upon only a segment of this period. Interestingly, in the middle of 2009, local planners and decision makers decided to make a change in the way the municipality was addressing its chronic flood problems. Most important was updating the jurisdiction's comprehensive plan, which set the stage for a far more innovative and aggressive local flood management program. The revised plan of 2009 (based on an Evaluation and Appraisal Report conducted in 2007) admits that "present ordinances do not adequately address natural hazard mitigation or land use in flood prone areas" (Coastal Management Element, St. Petersburg Comprehensive Plan, 2009). The plan then goes on to adopt a series of policies to better incorporate land use techniques into mitigation and address head-on the issue of repetitive flood loss properties.

Among other new policies, it seeks to expand the city's "green permeable open space so as to provide maximum area for shallow aquifer recharge and storm-water filtration," placing a greater emphasis on the role of natural functions in attenuating floods. The plan also requires the implementation of a Stormwater Management Master Plan, including the completion of 85 specific projects by the end of 2025. Policies adopted in the comprehensive plan were then integrated into St. Petersburg's stand-alone Floodplain Management Plan, also updated in 2009. Based on the city's new policies and standards to reduce the adverse impacts of floods, in the same year its CRS class improved to a 6, affording residents in the floodplain a 20% discount on their federal flood insurance premiums. Hopefully, all of the progress St. Petersburg has made with respect to flood planning and management will enable it to become a safer, more resilient community in the future.

Palm Beach Gardens, Florida

Palm Beach Gardens is situated on the southeastern side of the Florida peninsula in Palm Beach County. Almost all of the city's area lies slightly inland to the west of the Intracoastal Waterway, such that it has no direct shoreline on the Atlantic Ocean. However, Palm Beach Gardens' close proximity to the beach and easy access to all of south Florida's amenities has made it a rapidly growing, wealthy community. According to University of Florida's Bureau of Economic and Business Research (BEBR) data, the City's population increased 53% from 22 965 to 35 058 between 1990 and 2000. Palm Beach Gardens added another 14 329 persons in the following seven years, making the 2007 estimated population 49 387. All projections indicate the City will continue to experience rapid population growth; by 2015, the population is expected to reach 61 076 (another 23% increase).

With new residents came housing units, mostly in the form of single-family residential structures built in heavily landscaped gated communities. From 2000 to 2004, for example, Palm Beach Gardens added 6563 units to its housing stock through several major planned unit developments. In 2004, there was a total of 24 688 housing units, the vast majority (over 70%) single-family residential. Occupants of these dwellings are generally wealthy and well educated with much to lose in terms of flood damages. In 2007, the median household income was almost $60 000 and only 5.6% of residents were living below the poverty line. Based on 2006 property appraisal data, the average single home value swelled to $578 670, and the condominium average value reached $280 634.

Although Palm Beach Gardens has a large financial investment in the form of residential development and tax base potentially at risk to flooding, the city is located in a relatively secure area from a physical vulnerability standpoint. Less than 5% of the jurisdiction lies within the 100-year floodplain and most of its existing wetlands (in western portions of its boundaries) remain undeveloped. There are no riverine floodplains within Palm Beach Gardens, but the Loxahatchee Slough is a floodplain for sheet flow through naturally occurring wetlands. The west basins in the city typically drain to the west and follow the historic flow path to the Loxahatchee Slough and River. The middle and eastern portions of the City drain via human-made canals to the Intracoastal Waterway and/or Lake Worth Lagoon, then eventually into the Atlantic Ocean.

Furthermore, the Palm Beach Gardens' municipal boundaries lie west of the Intracoastal Waterway so there is no direct shoreline that could be easily inundated. The City's slightly interior position also makes for a small coastal planning area with few residential parcels for which flooding could be a concern. Low risk of exposure to flooding is one reason why Palm Beach Gardens incurred only minor property loss during our study period. The total insured loss from 1996 to 2007 was less than $1 million. This amounts to only about $26 per person during this time frame, which is even more notable considering Palm Beach County ranked third in all of Florida for flood losses since 1960 and in the top 10 of all 121 counties in our two-state sample.

Palm Beach County's low rate of flood damages is not simply a function of its position and geophysical characteristics, but also its local policy learning and response to the potential threat. The locality had the second highest overall mitigation score in our survey (see Chapter 5 for more details) and the top score for structural techniques. A tour of the city reveals an intricate network of drainage canals, detention ponds, and control structure that are so well embedded into the residential landscape they appear and act as natural amenities. For storm-water management, Palm

Beach Gardens also makes use of a high-capacity canal to the west operated by the South Florida Water Management District that drains the Loxahatchee Slough and the Intracoastal Waterway (Lake Worth Lagoon) to the east. As a whole, this multi-hydraulic classification drainage system (significantly enhanced in response to record rainfall in 1995) efficiently handles runoff caused by heavy subtropical rainfall events, even in areas dominated by impervious surfaces. Specific drainage projects are further specified in the City Stormwater Management Plan adopted in 2002.

Palm Beach Garden planners match structural approaches to flood mitigation with strong non-structural and land use planning practices. The local comprehensive plan, revised in 2008, establishes several "avoidance" strategies to keep residences and public buildings away from sensitive and vulnerable areas. For example, the plan designates an urban growth boundary, which bisects the city and seeks to retain high density/urban development to the east of Loxahatchee Slough. The UBG not only curtails sprawling growth, but prevents the most intense development and impervious surface from encroaching on floodplains and naturally occurring wetlands.

Another land use planning technique employed in the comprehensive plan is a density reduction overlay zone covering an area (in the central part of the municipality) susceptible to sheet flow flooding and associated wetlands. This zone is literally draped over existing regulations to require a 50% reduction in development density and a resulting gross density potential of two dwelling units per acre. While the overlay zone does not prohibit all development, it greatly reduces the potential exposure of residential structures to flood risk. Palm Beach Gardens also uses conservation zones and protected areas to minimize flood vulnerability. In fact, most of the western portion of the city is designated as either conservation or very low-density residential. In total, over 41% of Palm Beach Gardens is designated as conservation land use, which amounts to 317 acres per 1000 persons. Of course, what is important from a flood mitigation perspective is not how much land is protected, but where the protection occurs. In this sense, the City does an effective job of placing conservation zones where they are needed most to safeguard residents and critical natural resources.

Finally, the Palm Beach Gardens comprehensive plan contains a policy prohibiting public/institutional buildings within 100-year floodplains. This land use restriction ensures that critical facilities are kept out of flood-prone areas so that government services remain uninterrupted during flooding events. Through these planning techniques, the City as a whole is more resilient to the adverse economic and human safety impacts associated with floods (for more details, see Palm Beach Gardens Comprehensive Plan, 2008).

Palm Beach Gardens' commitment to flood mitigation is affirmed by its participation in FEMA's CRS. In 2008, around the time its revised comprehensive plan was adopted, the city was designated a class 7, allowing residents living in the floodplain a 15% discount on their federal insurance premiums. While local public officials intend to improve their CRS rating, Palm Beach Gardens already scored far above average (240 points where 90 is the national average) for its public outreach efforts (activity 330) to inform residents about flood risks and implementation plans. The city also scored above average for achieving higher regulatory standards (activity 430), such as its reduced density overlay as well as for its vigilance in drainage system maintenance (activity 540).

Overall, decision makers in Palm Beach Gardens have responded well to the threats associated with flooding. A well-designed storm-water drainage system combined with spatially targeted land use policies has kept recent flood losses to a minimum. Having a small population with large amounts of financial resources and relatively low level of physical risk has undoubtedly contributed to the city's success. It will be critical, however, for Palm Beach Gardens to maintain its current approach to development and mitigation as the population continues to expand.

Part IV

Policy implications and recommendations

11

Flood policy recommendations

The purpose of our study is to better understand the causes and consequences of flooding so that our findings can help inform local jurisdictions on how best to mitigate hazards at the local level. Based on the results of our empirical analyses, we present a series of policy recommendations as guidance to effectively mitigate the adverse impacts of floods. Decision makers, coastal stakeholders, and other interested parties within and outside of our study areas can use the following recommendations to enhance the flood resiliency of their communities.

Non-structural flood mitigation strategies are a viable alternative

Through both quantitative and qualitative evidence, we have shown the effectiveness of using non-structural mitigation techniques to reduce the adverse impacts of flooding events. The rush to build levees, dams, and other public works after a destructive flood should be met with caution. As already noted, these structures are not infallible, require considerable financial resources, and can induce a false sense of security leading to increased development in vulnerable areas. In contrast, non-structural, land use-based measures may significantly reduce observed property damages from floods by guiding development away from flood-prone areas. Time and time again, using multiple statistical procedures, we have shown the value of this approach for enhancing the resiliency of coastal communities. Based on our data, the most effective strategies include setbacks or buffers, "pocket" protected areas, strong construction and building codes, and specific flood policies embedded into local land use plans.

Flood regulations should exceed NFIP requirements

The National Flood Insurance Program, the cornerstone of floodplain management in the U.S., should only be considered a starting point for a local flood mitigation program. If a city or town is serious about protecting citizens and property, additional measures beyond basic floodplain regulations should be implemented. FEMA's CRS is one alternative, providing incentives for localities to adopt primarily non-structural mitigation strategies involving, among others, outreach and education, open space protection, and higher regulatory standards such as freeboard and reduced development densities. This program is particularly attractive because it rewards mitigation efforts by offering discounts on federal flood insurance premiums.

Our analyses demonstrate a direct relationship between higher CRS scores and reduced human casualties and property damage caused by floods. In Texas, a step up in the CRS scoring ladder translated into almost $39 000 of reduced damage per flood from 1997 to 2001 (Brody *et al.*, 2008b: 1–18). In Florida, this saving increased to over $303 000 (Brody *et al.*, 2007a: 330–345).

The FEMA CRS is just one program highlighted throughout the book for its effectiveness in safeguarding local communities from the dangers of flooding. At the federal level, this program is the most advanced, comprehensive, and successful in terms of helping localities reduce their exposure to flood risks. However, we find many jurisdictions that are nonparticipants in the CRS aggressively pursuing non-structural flood mitigation strategies. These strategies include strong flood mitigation elements within comprehensive plans, local mitigation strategy plans and watershed management plans, which are potentially effective policy vehicles if implemented at the local level.

Avoidance strategies should be considered the first approach to mitigation

Avoidance strategies or spatially guided development should be the first line of defense against flooding. This approach to flood mitigation involves keeping structures and their inhabitants away from the most flood-prone and ecologically vulnerable areas. The 100-year floodplain is typically considered the barometer for increasing risk of exposure. In the U.S., development in a floodplain often triggers additional regulations, but decision makers should question whether building in these areas should occur at all. Outwardly sprawling development from urban centers can infringe upon less expensive areas in the floodplain, which were once considered off limits. Homes and businesses in these areas are then subject to increased risk of flood loss or must shoulder the burden of expensive drainage or flood control projects.

Even within the 100-year floodplain, there are areas more vulnerable than others. Extremely low-lying parcels, creek-beds, and wetlands are all landscape features that pose additional risk if built upon. Our study shows that setbacks, buffers, and "pocket" protected areas are especially effective in reducing property damage from floods because these strategies remove people from the most vulnerable locations. For example, even when controlling for the amount of precipitation, area of floodplain, mean income, and number of structures within each jurisdiction in our Texas–Florida study areas, the implementation of protected areas for flood mitigation significantly reduces the observed amount of insured flood losses. In fact, from 2006 to 2007, jurisdictions with this mitigation strategy in place saved, on average, $298 965 in property damages. Setbacks and buffers were also quite effective during the same time period, producing an average savings in insured flood damages of $199 148 per jurisdiction.

The supporting role of retention and detention should be strongly considered

When avoidance strategies are not feasible or desired, the role of retention and detention ponds to collect, hold, and slowly release runoff become particularly important at the site level. In a suburban landscape, these devices should be integrated into the residential community, rather than constructed as a hidden ditch behind buildings. We found that many developments throughout Florida use retention ponds (which hold a certain amount of water indefinitely) as natural amenities with ringing paths, benches for resting, and places for recreation. These amenities can increase the value of surrounding homes and help build a sense of community among residents. If constructed properly with native planting, these detention ponds can also support wildlife habitat.

Naturally occurring wetlands should be considered a flood control device

Our study is one of the first to empirically demonstrate the value of naturally occurring wetlands for reducing flooding and flood damages over a large study area (Chapter 6). In both Texas and Florida, our models show how the removal or alteration of wetlands compromises their overall capacity to capture, store, and slowly release water runoff. The consequences of building roads, parking lots, and houses in these sensitive areas are increased amounts of flooding and associated property damage. For example, each wetland permit issued by the USACE in Florida from 1997 to 2001 increased the average cost of each flood by over $989. This wetland "permit effect" equates to, on average, $563 451 of flood damage per county per year, which averages out to about $30 426 354 per year for the entire state (Brody *et al.*, 2007a: 330–345).

Naturally occurring wetlands should be valued not only for wildlife habitat, fisheries, and recreation but also for flood control. In this sense, wetland protection for flood management should be more systematically integrated into local plans and zoning ordinances. Once identified by a locality, these critical areas (often but not always in the floodplain) can be protected from development through multiple planning techniques, including zoning restrictions, overlay zones, land acquisition programs, clustered development, density bonuses, and transfer of development rights (Brody and Highfield, 2005: 159–175). The goal of local decision makers in this case should be to allow development to proceed without compromising the hydrological function of wetland systems. This approach to wetland protection does occur, but only sporadically throughout the study area.

A second recommendation about wetland protection is the idea of internalizing the cost of wetland alteration in the development process. If we have the data and analytical techniques to literally cost out the impact of a wetland permit in terms of future property damage caused by floods, should this cost not be borne by the applicant? Most permits issued by the USACE, including letters of permission, nationwide, and general permits, currently have no fee. Individual permits cost only $10 for individuals and $100 for commercial projects (Highfield and Brody, 2006: 23–30). Localities should consider setting the price of a wetland permit at an appropriate level (in our case, $989) to more fully internalize the true cost and reduce the attractiveness of altering wetlands in the first place.

Policies should be embedded into the local regulatory framework

Many local jurisdictions rely solely on stand-alone plans, such as local mitigation strategies, drainage basin plans, floodplain management plans, and storm-water management plans, to mitigate their flood problems. While these instruments may be technically competent, unless the policies they contain are embedded into the local regulatory framework, they will not live up to their potential. Development decisions at the community level are made and implemented through local land use plans, zoning ordinances, and development codes. These are the tools through which flood management must ultimately be executed, even if stand-alone plans are already in place.

Indeed, our results demonstrate the effectiveness of having flood policies incorporated into local land use plans. In Chapter 7, we show through correlation analysis that when a jurisdiction adopts a stand-alone flood plan there is no appreciable reduction in flood damages. However, a single flood policy integrated into a local land use plan has a statistically significant effect on reduced property damage from floods. When we control for floodplain area, median income, and the amount of precipitation in each jurisdiction, we find that having a flood management policy in a local land use plan reduces observed damage an average of $324 772. In contrast, a stand-alone plan under the same modeling conditions can save a jurisdiction an average of $34 000 in insured property damage caused by floods.

No single flood mitigation strategy is sufficient

Throughout this study we have evaluated specific flood policies independently to gauge their statistical effect on recorded property damages. However, in reality, a locality should adopt a combination of strategies, tools, and techniques to most effectively address flood issues. There is most likely no single recipe for constructing a flood management program. Each jurisdiction will have its own unique geophysical conditions, socioeconomic characteristics, and degree of political will. However, we can offer the following insights. First, there is a synergistic effect between flood policies, such that a combination will have an amplified effect in protecting a community. Layering multiple policies around a highly vulnerable area can provide blanket protection that a single strategy would be unable to accomplish. For example, a "pocket" protected area encompassing wetlands, surrounding by a buffer of restricted use, surrounded by an area of reduced density and strong building codes could be more effective in reducing flood losses than implementing just one of these policies.

Second, assuming development will continue in low-lying coastal areas, a hybrid approach between structural and non-structural flood mitigation alternatives will be necessary. The debate on mitigation is often structural versus non-structural, but we advocate a balanced strategy where the two approaches work in concert (as is often done in the Netherlands) to create communities that are resilient to floods. For instance, incorporating a series of well-placed detention ponds or levees into the example above could further reduce the level of exposure to flood risks and the amount of property loss.

The 100-year floodplain should not be the sole driver of management decisions

Throughout this book, we have used the FEMA-derived 100-year floodplain as a key indicator of flood risk. We also note that localities use these boundaries as the basis for their flood mitigation programs as required by the federal government. Using floodplains to inform plans and policies is indeed useful, but we urge caution to decision makers who are fixated on floodplain boundaries when assessing and mitigating risk of exposure. Floodplain boundaries are often dated (the national dataset is in the process of being updated) and can be inaccurate, especially if adjacent development alters the structure of the hydrological system. Too many times we hear reports of residences outside the designated floodplain boundaries that have never experienced flooding being inundated after adjacent neighborhoods are developed or following a previously unprecedented rainfall event. In fact, of the $286 million in repetitive flood losses experienced in Harris County, Texas, (an area known for sprawling development patterns and abundance of impervious surfaces) since 1979, over 47% of property damages were actually outside of the 100-year floodplain.

Another problem with relying on the FEMA floodplain designation to assess and prepare for risk is that it acts as the sole trigger for mandatory purchase of flood insurance. Residential structures within the floodplain are required to obtain federal insurance if applying for a mortgage because of the heightened risk. But what about homeowners who might be located just 1 ft (0.30 m) outside this demarcation? Residents investing in properties any distance (however short) outside of the floodplain are neither required nor encouraged to buy flood insurance. In fact, once in the "X" zone, residents may not even be aware of their home's proximity to the nearest floodplain boundary. As a result, property owners could be caught off-guard and uninsured when rising waters invade their homes.

Public organizations at all levels must be more aggressive in informing their constituents not only if their homes are located in or out of a floodplain, but how far they are from the nearest floodplain boundary. The data and tools already exist to determine where a residential structure is located in relation to multiple flood risk factors and deliver this information to the public over the internet. Enabling residents in flood-prone areas to make more informed decisions about their relative level of exposure and the mechanisms available for mitigation needs to be made a higher priority.

Local and federal decision makers should consider establishing a more sophisticated barometer for determining a property's risk for flooding. Given the potential inaccuracies and changing nature of floodplain boundaries, a buffer approach could be taken where residents within a certain proximity of a floodplain are at least notified of their risk and strongly encouraged to purchase flood insurance. The cost of flood insurance for a four-bedroom house outside (by even 1 ft (0.30 m)) a floodplain in Galveston County, Texas, one of the most vulnerable jurisdictions in our study area, is only about $355 per year. The fact is that no empirical study has been done to understand flood impacts for structures outside of the floodplain or the motivation of these homeowners to protect themselves on their own initiative.

Strong organizational capacity is a key aspect of flood management

Simply adopting a series of flood mitigation policies is usually not enough to safeguard a flood-prone community. These policies must instead be driven and backed by organizations with the capacity to ensure that flood mitigation techniques are properly implemented. In Chapter 7, for example, our data show that flood management programs are brought about by public organizations infused with characteristics of commitment, information sharing, leadership, staff expertise, and financial resources. These attributes help build enduring "policy legacies" because they ensure that mitigation strategies are implemented, monitored, and adjusted over time as new conditions arise. Strong capacity helps organizations adapt and learn from one flood event to the next so that over time, a community becomes more resilient.

We also demonstrate that strong organizations directly influence the amount of observed property damage from floods at the local level. Specifically, the commitment of publicly elected officials (e.g., zoning board members, mayor, judge, city councilmen) and the resources to maintain and retain a knowledgeable staff are critical to decreasing flood damages. Given these findings, we urge local jurisdictions to pay as much attention to developing organizations as they do to developing plans and policies. It is perhaps the combination of these elements that creates a pathway for local community flood resiliency.

Education and information dissemination is needed to inform the public

The decision to accept a certain level of flood risk is ultimately up to the homeowner, resident, business person, or investor. However, if these individuals and households are not fully aware of their risk, the planning and development system begins to break down. It is the responsibility of local government and the real-estate industry to properly inform residents of their exposure to flooding and provide information and training on how best to mitigate this risk. Too often we hear that homeowners did not know they were living in a floodplain, assumed by purchasing insurance they were protected, or thought they would never experience flood damage.

The data and communication techniques already exist to ensure that residents in areas vulnerable to flooding are apprised of potential risks before they make an investment. Indeed, educational outreach and information dissemination activities provide the majority of the point base for CRS-participating communities. Web-based GIS programs, public workshops, distributed guidebooks, and mail inserts are just a few of the techniques that should be implemented systematically among local communities. If individuals are fully informed when deciding where to live, work, or play, the desire to invest in flood-prone areas may substantially decline.

Improve available data on floods and flood damage in the U.S.

In Chapters 2 and 3, we organized data from multiple sources and time periods to paint an overall picture of the impacts of floods at various scales in the U.S.

As already noted, this was an arduous task because data are scattered across different organizations, collected using different methods and spatial units, and in some cases, not freely available to the public. Local jurisdictions encounter similar problems when trying to understand the degree to which flooding affects residents because conducting their own damage assessments is often not feasible. Localities cannot effectively mitigate flood impacts when they do not have a clear picture of exactly where and how much damage is occurring over time or if they receive conflicting evidence.

A central clearinghouse of data on the status and trends associated with flooding and flood damages is critical if the nation as a whole is going to improve its ability to reduce adverse impacts from these natural hazard events. This database must be web-accessible, easily incorporated into a GIS, address socioeconomic impacts, and be made available at local scales where the data are most needed. Access to high-quality flood data should not be restricted to large insurance companies or a select few in federal agencies. If successful mitigation requires implementing plans and policies at the local level (a central argument of this book), then detailed information about flood events must be made available from a single source at the smallest unit possible.

SHELDUS, which is produced by the Hazard Research Lab at the University of South Carolina (available at www.sheldus.org), is currently the best source of flood data because a user can search events across multiple attributes using a web interface. However, SHELDUS is not exclusive to flood hazards, is limited to the county level, and is not well known beyond academic circles. Given the tremendous economic impact of flooding in the U.S. alone, resources should be dedicated to establishing a flood data clearinghouse that collects, standardizes, maps, and disseminates information to local stakeholders.

12

Conclusions

Throughout this book, we have examined the impacts of floods in the U.S. from the national to the neighborhood scale in an effort to better elucidate the degree to which this natural hazard may impact economies, property, and daily lives. Using empirical data, we have investigated the major triggers of flooding, highlighting the importance of naturally occurring wetlands and other non-structural strategies in reducing flood loss. Most importantly, we have identified pathways for communities to mitigate the adverse impacts of chronic and persistent floods, particularly in the most vulnerable areas. Statistical analyses of observational data over multiple time periods and spatial scales have demonstrated which mitigation strategies are most effective in reducing damages, the purpose of which is to highlight policy alternatives that enhance a community's resiliency to flooding.

However, as already noted, there exists no magic remedy for effective flood mitigation that can be applied to every affected community. Each locality has its own set of geophysical, socioeconomic, and political characteristics that must be taken into account. The fact is, as explored in Chapter 9, flood mitigation is a long-term process during which jurisdictions and organizations must constantly adjust to ever-changing conditions and new streams of information. Communities must be flexible enough to learn from their mistakes and chart new policy courses accordingly. Our hope is that decision makers can use the evidence presented in this book to *establish enduring planning and policy legacies, so that over time, the adverse impacts of floods will diminish in their communities.*

Although this book is based on six years of evidence-based research on the topic of flooding, flood impacts and policy implications, it should be considered only a starting point for further investigation. The fact is, our analyses focuses only on two states. Future research should be national in scope and sub-local in scale to more fully understand what we know to be the costliest natural hazard in the country. In this sense, we should not be studying the record of flood loss only for counties, but for parcels and structures as well, so that we can make more spatially specific recommendations on mitigation strategies. Also, more work must be done on the effectiveness of specific mitigation activities to better discern the combination of techniques that offers the most potent defense against floods. The data and analytical methods to conduct this line of research already exist. What is missing is a full understanding of the severity of floods in the U.S. and the political will to investigate how to most effectively minimize their impacts.

References

Abell, RA (1999) *Freshwater Ecoregions of North America: A Conservation Assessment*, Island Press: Washington, DC.

Alexander, D (1993) *Natural Disasters*, Chapman & Hall: New York.

Alig, RJ, JD Kline, and M Lichtenstein (2004) "Urbanization on the US landscape: looking ahead in the 21st century," *Landscape and Urban Planning*, **69**, 219–234.

Ammon, DC, HC Wayne, and JP Hearney (1981) "Wetlands' use for water management in Florida," *Journal of Water Resources Planning and Management*, **107**, 315–327.

Arnold, CL and JC Gibbons (1996) "Impervious surface coverage: the emergence of a key environmental indicator," *Journal of the American Planning Association*, **62** (2), 243–258.

Ashley, ST and WS Ashley (2008) "Flood fatalities in the United States," *Journal of Applied Meteorology and Climatology*, **47**, 805–818.

Association of State Floodplain Managers (2000) *National Flood Programs in Review – 2000*, ASFPM: Madison, WI.

Association of State Floodplain Managers (2004) *No Adverse Impact Floodplain Management: Community Case Studies*. ASFPM: Madison, WI

Bagstad, KJ, K Stapleton, and JR D'Agostino (2007) Taxes, subsidies, and insurance as drivers of United States coastal development. *Ecological Economics*, **63**, 285–298.

Beatley, T (2000) "Preserving biodiversity: challenges for planners," *Journal of the American Planning Association*, **66** (1), 5–20.

Beatley, T and K Manning (1997) *Ecology of Place: Planning for Environment, Economy, and Community*, Island Press: Washington, DC.

Beem, B (2006) "Planning to learn: blue crab policymaking in the Chesapeake Bay," *Coastal Management*, **34** (2), 167–182.

Benfield, FK, MD Raimi, and DT Chen (1999) *Once there were greenfields: how urban sprawl is undermining America's environment, economy and social fabric*, Natural Defense Resource Council and Surface Transportation Policy Project.

Berke, F (2007) "Understanding uncertainty and reducing vulnerability: lessons from resilience thinking," *Natural Hazards*, **41** (2), 283–295.

Berke, P, D Roenigk, EJ Kaiser, and RJ Burby (1996) "Enhancing plan quality: evaluating the role of state planning mandates for natural hazard mitigation," *Journal of Environmental Planning and Management*, **39**, 79–96.

Berke, P, Y Song, and M Stevens (2009) "Integrating hazard mitigation into new urban and conventional developments," *Journal of Planning Education and Research*, **28** (4), 441–455.

Berke, PR and SP French (1994) "The influence of state planning mandates on local plan quality," *Journal of Planning Education and Research*, **13** (4), 237–250.

Birkland, TA, RJ Burby, D Conrad, H Cortner, and WK Michener (2003) "River ecology and flood hazard mitigation," *Natural Hazards Review*, **4** (1), 46–54.

Bras, RL (1990) *Hydrology: An Introduction to Hydrologic Science*, Addison-Wesley: Reading, MA.

Brezonik, PL and TH Stadelman (2002) "Analysis and predictive models of stormwater runoff volumes, loads and pollutant concentrations from watersheds in the twin cities metropolitan area, Minnesota, USA," *Water Resources*, **36**, 1743–1757.

Brody, SD (2003a) "Measuring the effects of stakeholder participation on the quality of local plans based on the principles of collaborative ecosystem management," *Journal of Planning Education and Research*, **22** (4), 107–119.

Brody, SD (2003b) "Are we learning to make better plans? A longitudinal analysis of plan quality associated with natural hazards," *Journal of Planning Education and Research*, **23** (2), 191–201.

Brody, SD (2003c) "Examining the effects of biodiversity on the ability of local plans to manage ecological systems," *Journal of Environmental Planning and Management*, **46** (6), 733–754.

Brody, SD (2008) *Ecosystem Planning in Florida: Solving Regional Problems through Local Decision Making*, Ashgate Press: Aldershot, UK.

Brody, SD and WE Highfield (2005) "Does planning work? Testing the implementation of local environmental planning in Florida," *Journal of the American Planning Association*, **71** (2), 159–175.

Brody, SD, V Carrasco, and WE Highfield (2006a) "Measuring the adoption of local sprawl reduction planning policies in Florida," *Journal of Planning Education and Research*, **25**, 294–310.

Brody, SD, WE Highfield, and S Thornton (2006b) "Planning at the urban fringe: an examination of the factors influencing nonconforming development patterns in southern Florida," *Environment and Planning B*, **33**, 75–96.

Brody, SD, S Zahran, P Maghelal, H Grover, and W Highfield (2007a) "The rising costs of floods: examining the impact of planning and development decisions on property damage in Florida," *Journal of the American Planning Association*, **73** (3), 330–345.

Brody, SD, WE Highfield, HC Ryu, and L Spanel-Weber (2007b) "Examining the relationship between wetland alteration and watershed flooding in Texas and Florida," *Natural Hazards*, **40** (2), 413–428.

Brody, SD, SE Davis III, WE Highfield, and S Bernhardt (2008a) "A spatial–temporal analysis of wetland alteration in Texas and Florida: thirteen years of impact along the coast," *Wetlands*, **28** (1), 107–116.

Brody, SD, S Zahran, WE Highfield, H Grover, and A Vedlitz (2008b) "Identifying the impact of the built environment on flood damage in Texas," *Disasters*, **32** (1), 1–18.

Brody, SD, S Zahran, SP Bernhardt, and JE Kang (2009a) "Evaluating local flood mitigation strategies in Texas and Florida," *Built Environment*, **35** (4), 492–515.

Brody, SD, JE Kang, and SP Bernhardt (2009b) "Identifying factors influencing flood mitigation at the local level in Texas and Florida: the role of organizational capacity," *Natural Hazards*, **52** (1), 167–184.

Brody, SD, S Zahran, WE Highfield, S Bernhardt, and A Vedlitz (2009c) "Policy learning for flood mitigation: a longitudinal assessment of CRS activities in Florida," *Risk Analysis*, **29** (6), 912–929.

Bullock, A and M Acreman (2003) "The role of wetlands in the hydrological cycle," *Hydrology and Earth System Sciences*, **7** (3), 358–389.

Burby, RJ (2003) "Making plans that matter – citizen involvement and government action," *Journal of the American Planning Association*, **69** (1), 33–49.

Burby, RJ (2005) "Have state comprehensive planning mandates reduced insured losses from natural disasters?," *Natural Hazards Review*, **6** (2), 67–81.

Burby, RJ and LC Dalton (1994) "Plan can matter! The role of land use plans and state planning mandates in limiting the development of hazardous areas," *Planning Administration Review*, **54** (3), 229–238.

Burby, R J, and P May, with P Berke, L Dalton, S French, and E Kaiser (1997) *Making Governments Plan: State Experiments in Managing Land Use*, Johns Hopkins University Press: Baltimore, MD.

Burby, RJ and PJ May (1998) "Intergovernmental environmental planning: addressing the commitment conundrum," *Journal of Environmental Planning and Management*, **41** (1), 95–110.

Burby, RJ, SP French, B Cigler, EJ Kaiser, D Moreau, and B Stiftel (1985) *Flood Plain Land Use Management: A National Assessment*, Westview Press: Boulder, CO.

Burby, RJ, S Bollens, J Haloway, EJ Kaiser, D Mullan, and J Schafer (1988) *Cities Under Water: A Comparative Evaluation of Ten Cities' Efforts to Manage Floodplain Land Use*, Institute of Behavioral Science, University of Colorado: Boulder, CO.

Burby, RJ, T Beatley, PR Berke, RE Deyle, SP French, DR Godschalk, EJ Kaiser, JD Kartez, PJ May, R Olshansky, RG Paterson, and RH Platt (1999) "Unleashing the power of planning to create disaster-resistant communities," *Journal of the American Planning Association*, **5** (3), 247–258.

Burchell, RW, NA Shad, D Listokin, H Phillips, A Downs, S Seskin, JS Davis, T Moore, D Helton, and M Gall (1998) *The Costs of Sprawl – Revisited*. Report 39, Transit Cooperative Research Program, Transportation Research Board, National Academy Press: Washington DC.

Burges, SJ, MS Wigmosta, and JM Meena (1998) "Hydrological effects of land-use change in a zero-order catchment". *Journal of Hydrological Engineering*, **3**, 86–97.

Burns, D, T Vitvar, J McDonnell, J Hassett, J Duncan, and C Kendall (2005) "Effects of suburban development on runoff generation in the Croton River basin, New York, USA," *Journal of Hydrology*, **311**, 266–281.

Campbell, DA, CA Cole, and RP Brooks, (2002) "A comparison of created and natural wetlands in Pennsylvania, USA," *Wetlands Ecology and Management*, **10**, 41–49.

Carter, RW (1961) "Magnitude and frequency of floods in suburban areas," In *Short Papers in Geologic and Hydrologic Sciences*, USGS Professional Paper 424-B: B9–B11. U.S. Geological Service: Washington, DC.

Chaptin, T, C Connerly, and H Higgins, eds (2007) *Growth Management in Florida: Planning for Paradise*. Ashgate Press: Burlington, VT.

Cole, CA and RP Brooks (2000) "A comparison of the hydrologic characteristics of natural and created mainstem floodplain wetlands in Pennsylvania," *Ecological Engineering*, **14**, 221–231.

Cole, CA and D Shafer, (2002) " Section 404 wetland mitigation and permit success criteria in Pennsylvania, USA, 1986–1999," *Environmental Management*, **30** (4), 508–515.

Congressional Budget Office (2009) *The National Flood insurance Program: Factors Affecting Actuarial Soundness*, The Congress of the United States, Congressional Budget Office, Pub. No. 4008.

Connally, KD, SM Johnson, and DR Williams (2005) *Wetlands Law and Policy: Understanding Section 404*, American Bar Association: Chicago, IL.

Crossett, KM, TJ Culliton, PC Wiley, and TR Goodspeed (2004) *Population Trends Along the Coastal United States*, National Oceanic and Atmospheric Administration: Silver Spring, MD.

Dahl, TE (2000) *Status and Trends of Wetlands in the Conterminous United States 1986 to 1997*, U.S. Department of the Interior, Fish and Wildlife Service: Washington, DC.

Dahl, TE (2006) *Status and Trends of Wetlands in the Conterminous United States 1998 to 2004*, U.S. Department of the Interior Fish and Wildlife Service: Washington, DC.

Dahl, TE and CE Johnson (1991) *Status and Trends of Wetlands in the Conterminous United States, mid-1970s to mid-1980s*, U.S. Department of the Interior, Fish and Wildlife Service: Washington, DC.

Dalton, LC and RJ Burby (1994) "Mandates, plans and planners: building local commitment to development management," *Journal of the American Planning Association*, **60** (4), 444–461.

Daniel, C (1981) Hydrology, geology, and soils of pocosins: a comparison of natural and altered systems, In *Pocosins: A Conference on Alternate Uses of the Coastal Plain Freshwater Wetlands of North Carolina*, ed. CJ Richardson, Hutchinson Ross Publishing Company: Stroudsburg, PA, pp. 69–108.

Dennison, MS and JF Berry (1993) *Wetlands: Guide to Science, Law, and Technology*, Noyes Publications: Park Ridge, NJ.

Downing, DM, C Winer, and LD Wood (2003) "Navigating through clean water act jurisdiction: a legal review," *Wetlands*, **23** (3), 475–493.

Deyle, RE, TS Chapin, and EJ Baker (2008) The proof of the planning is in the platting: an evaluation of Florida's hurricane exposure mitigation planning mandate. *Journal of the American Planning Association*, **74** (3), 349–370.

Espey, WH, CW Morgan, and FD Masch (1965) *A Study of Some Effects of Urbanization on Storm Runoff From a Small Watershed, Tech. Rep. 44D 07–6501 CRWR-2*, University of Texas, Center for Research in Water Resources: Austin, TX.

Federal Emergency Management Agency (1997) *MultiHazard Identification and Risk Assessment*, Washington, DC: U.S. Government Printing Office.

Federal Emergency Management Agency (2002) *National Flood Insurance Program Description*, FEMA, Federal Insurance and Mitigation Administration: Washington, DC.

Federal Emergency Management Agency (2007a) *The National Flood Insurance Program*, available at: www.fema.gov/about/programs/nfip/index.shtm (accessed December 27, 2008).

Federal Emergency Management Agency (2007b) *National Flood Insurance Program Community Rating System Coordinator's Manual*, available at: http://training.fema. gov/EMIWeb/CRS/2007%20CRS%20Coord%20Manual%20Entire.pdf (accessed July 1, 2010).

Federal Emergency Management Association (2008) *Mitigation Best Practices*, Region VI, November.

Federal Emergency Management Agency (2010) *Policy Statistics*, available at: http://bsa. nfipstat.com/reports/1011.htm (accessed May 20, 2010).

Florida Division of Emergency Management (2008) *2008 DEM sunset review report*, available at: www.floridadisaster.org/documents/2008%20DEM%20Sunset%20Review%20 Report%20FINAL.pdf (accessed February 20, 2008).

Folke, C, T Hahn, P Olsson, and J Norberg (2005) "Adaptive governance of social–ecological systems," *Annual Review of Environment & Resources*, **30** (1), 441.

Gall M, AB Kevin, and SL Cutter (2009) "When do losses count? Six fallacies of natural hazards loss data," *Journal of the American Meteorological Society*, **90** (6), 799–809.

Gallihugh, JL and JD Rogner (1998) *Wetland Mitigation and 404 Permit Compliance Study, Volume 2*, U. S. Fish and Wildlife Service: Barrington, IL.

Godschalk, DR, DJ Brower, and T Beatley (1989) *Catastrophic Coastal Storms: Hazard Mitigation and Development Management*, Duke University Press: Durham, NC.

Godschalk, DR, T Beatley, P Berke, D Brower, and EJ Kaiser (1999) *Natural Hazard Mitigation: Recasting Disaster Policy and Planning*, Island Press: Washington, DC.

Government Accountability Office (2005) *Wetlands Protection: Corps of Engineers Does Not Have an Effective Oversight Approach to Ensure That Compensatory Mitigation Is Occurring*, available at: www.gao.gov/products/GAO-05–898 (accessed March 20, 2008).

Grindle MS and ME Hilderbrand (1995) Building sustainable capacity in the public sector: what can be done?, *Public Administration and Development*, **15**, 441–463.

Hall, PA (1993) "Policy paradigms, social-learning, and the state: the case of economic policy-making in Britain," *Comparative Politics*, **25** (3), 275–296.

Handmer, J (1996) "Policy design and local attributes for flood hazard management," *Journal of Contingencies and Crisis Management*, **4** (4), 189–197.

Hartig JH, NL Law, D Epstein, K Fuller, J Letterhos, and G Krantzberg (1995) "Capacity building for restoring degraded areas in the Great Lakes", *International Journal of Sustainable Development and World Ecology*, **2**, 1–10.

Hartvelt, F and DA Okun (1991) "Capacity building for water resources management," *Water International*, **16**, 176–183.

Hazards Research Lab (SHELDUS) (2008) *The spatial hazard events and losses database for the United States, Version 4.1.*, available at: http://webra.cas.sc.edu/hvri/products/ sheldus.aspx (accessed January 13, 2008).

Hazards and Vulnerability Research Institute, University of South Carolina (2009) *Oxfam America*, available at: http://adapt.oxfamamerica.org/ (accessed April 10, 2009).

Heclo, H (1974) *Modern Social Politics in Britain and Sweden; From Relief to Income Maintenance*, Yale Studies in Political Science, Yale University Press: New Haven, CT.

Heikuranen, L (1976) *Comparison Between Runoff Condition on a Virgin Peatland and a Forest Drainage Area*, Proceedings of the Fifth International Peat Congress, pp. 76–86.

Heinz Center (2000) *The Hidden Costs of Coastal Hazards: Implications for Risk Assessment and Mitigation*, Island Press: Washington, DC.

Hey, DL (2002) "Modern drainage design: the pros, the cons, and the future," *Hydrological Science and Technology*, **18** (14), 89–99.

Highfield, WE (2008) *Longitudinal Analysis of the Relationship between Section 404 Permit Activity and Peak Streamflow in Coastal Texas*, Unpublished Ph.D. thesis, Texas A&M University: College Station, TX.

Highfield, WE and SD Brody (2006) "The price of permits: measuring the economic impacts of wetland development on flood damages in Florida," *Natural Hazards Review*, **7** (3), 23–30.

Hirsch, RM, JF Walker, JC Day, and R Kallio (1990) "The influence of man on hydrological systems," In *Surface Water Hydrology*, eds MG Wolman and HC Riggs, Geological Society of America: Boulder, CO, **0–1**, pp. 329–359.

Hirschhorn, JS (2001) "Environment, quality of life, and urban growth in the new economy," *Environmental Quality Management*, **10** (3), 1–8.

Hoch, CJ, LC Dalton, and FS So, eds (2000), *The Practice of Local Government Planning* (3rd edition). International City/County Management Association: Washington, DC.

Holling, CS (1995) "What barriers? What bridges?" In *Barriers and Bridges to the Renewal of Ecosystems and Institutions*, eds LH Gunderson, CS Holling, and SS Light, Columbia University Press: New York, pp. 3–34.

Holling, CS (1996) "Surprise for science, resilience for ecosystems, and incentives for people," *Ecological Applications*, **6** (3), 733–735.

Holling, CS and United Nations Environment Programme (1978) *Adaptive Environmental Assessment and Management*. International Series on Applied Systems Analysis: Laxenburg, Austria.

Holway, JM and RJ Burby (1990) The effects of floodplain development controls on residential land values. *Land Economics*, **66** (3), 259–271.

Holway, JM and RJ Burby (1993) Reducing flood losses through local planning and land use controls. *Journal of the American Planning Association*, **59** (3), 205–216.

Honadle, BW (1981) "A capacity-building framework: a search for concept and purpose," *Public Administration Review*, **43** (5), 575–580.

Horton, RE (1932) "Drainage basin characteristics," *Transactions of the American Geophysical Union*, **13**, 350–361.

Hsu, MH, SH Chen, and TJ Chang (2000) "Inundation simulation for urban drainage basin with storm sewer system," *Journal of Hydrology*, **234**, 21–37.

Innes, J (1996) "Planning through consensus building: a new view of the comprehensive planning ideal," *Journal of the American Planning Association*, **62**, 460–472.

Interagency Floodplain Management Review Committee (1994) *Sharing the Challenge: Floodplain Management into the 21st Century* Executive Office of the President: Washington, DC.

Issuance of Nationwide Permits (2005) "Notice," *Federal Register*, **67** (10), 2019–2095.

Ivey, JL, RC Loe, and RD Kreutzwiser (2002) "Groundwater management by watershed agencies: an evaluation of the capacity of Ontario's conservation authorities," *Journal of Environmental Management*, **64**, 311–331.

Johnson, CL, SM Tunstall, and EC Penning-Rowsell (2005) "Floods as catalysts for policy change: historical lessons from England and Wales," *International Journal of Water Resources Development*, **21** (4), 561–575.

Johnston, CA, NE Detenbeck, and GJ Niemi (1990) "The cumulative effect of wetlands on stream water quality and quantity: a landscape approach," *Biogeochemistry*, **10** (2), 105–141.

Kahn, ME (2000) "The environmental impact of suburbanization," *Journal of Policy Analysis and Management*, **19** (4), 569–586.

Kang, JE (2009) *Mitigation Flood Loss Through Local Comprehensive Planning in Florida*, Ph.D. thesis, Texas A&M University: College Station, TX.

Kelly, NM (2001) "Changes to the landscape pattern of coastal North Carolina wetlands under the clean water act, 1984–1992," *Landscape Ecology*, **16** (1), 3–16.

Kentula, ME, JC Sifneos, JW Good, M Rylko, and K Kunz (1992) "Trends and patterns in section 404 permitting requiring compensatory mitigation in Oregon and Washington, USA," *Environmental Management*, **16** (1), 109–119.

Kenworthy, JR and FB Laube (1999) "Patterns of automobile dependence in cities: an international overview of key physical and economic dimensions with some implications for urban policy," *Transportation Research Part A*, **33**, 691–723.

King, RO (2008) *National Flood Insurance Program:Treasury Borrowing in the Aftermath of Hurricane Katrina*. CRS Report for Congress, Order Code RS22394. Congressional Research Service: Washington, DC.

Kunreuther, H, J Richard, and S Roth, eds (1998) *Paying the Price: The Status and Role of Insurance Against Natural Disasters in the United States*. Joseph Henry Press: Washington, DC.

Kusler, JA and ME Kentula (1990) *Wetland Creation and Restoration – The Status of the Science*, Island Press: Washington, DC.

Lang, RE (2003) *Beyond Edge City: Office Sprawl in South Florida*, Center on Urban and Metropolitan Policy-Survey Series, The Brookings Institute: Washington, DC.

Larson, L and D Pasencia (2001) "No adverse impact: new direction in floodplain management policy," *Natural Hazards Review*, **2** (4), 167–181.

Laurian, L, M Day, M Backhurst, P Berke, N Ericksen, J Crawford, J Dixon, and S Chapman (2004) "What drives plan implementation? Plans, planning agencies and developers," *Journal of Environmental Planning and Management*, **47** (4), 555–577.

Lee, KN (1992) "Ecologically effective social organization as a requirement for sustaining watershed ecosystems," In *Watershed Management: Balancing Sustainability and Environmental Change*, ed. RJ Naiman, Springer-Verlag: New York, pp. 542.

Lee, KN (1993) *Compass and Gyroscope: Integrating Science and Politics for the Environment*, Island Press: Washington, DC.

Leopold, LB (1994) *A View of the River*, Harvard University Press: Cambridge, MA.

Lewis, WM (2001) *Wetlands Explained: Wetland Science, Policy, and Politics in America*, Oxford University Press: New York, NY.

Luloff, AE and KP Wilkinson (1979) "Participation in the national flood insurance program: a study of community attentiveness," *Rural Sociology*, **44**, 137–152.

Matthai, HF (1990) "Floods," In *The Geology of North America Vol. 0–1, Surface Water Hydrology*, eds MG Wolman and HC Riggs, The Geologic Society of America: Boulder, CO, pp. 97–120.

May, PJ (1992) "Reconsidering policy design: policies and publics," *Journal of Public Policy*, **11** (2), 187–206.

May, PJ (1998) *Fostering Policy Learning: A Challenge for Public Administrators*, Department of Political Science, University of Washington: Seattle, WA.

Mileti, DS (1999) *Disaster by Design*, Joseph Henry Press: Washington, DC.

Miller, VC (1953) *A Quantitative Geomorphic Study of Drainage Basin Characteristics in the Clinch Mountain Area, Virginia and Tennessee, Technical Report 3*, Department of Geology, Columbia University: New York, NY.

Mitch, WJ and JG Gosselink (2000) *Wetlands* (3rd edition), John Wiley & Sons: New York, NY.

Moore, DE and RL Cantrell (1976) "Community response to external demands: an analysis of participation in the federal flood insurance program," *Rural Sociology*, **41**, 484–508.

National Flood Insurance Program (2007) available at: http://www.fema.gov/business/nfip/ (accessed November 17, 2010).

National Research Council (2000) *Risk Analysis and Uncertainty in Flood Damage Reduction Studies*, National Academy Press: Washington, DC.

National Research Council (2001) *Compensating for Wetland Losses Under the Clean Water Act*, National Academy Press: Washington, DC.

National Weather Service (2004) *Flood Losses: Compilation of Flood Loss Statistics*, available at: www.nws.noaa.gov/hic/flood_stats/Flood_loss_time_series.shtml (accessed December 22, 2008).

Nelson, AC and SP French (2002) "Plan quality and mitigating damage from natural disasters: a case study of the Northridge earthquake with planning policy considerations," *Journal of the American Planning Association*, **68** (2), 194–207.

Novitski, RP (1979) "Hydrologic characteristics of Wisconsin's wetlands and their influence on floods, streamflow and sediment," In *Wetland Functions and Values: the State of our Understanding*, eds PE Greeson, JR Clark, and JE Clark, American Water Resources Association: Minneapolis, MN, pp. 377–388.

Novitski, RP (1985) *The Effects of Lakes and Wetlands on Flood Flows and Base Flows in Selected Northern and Eastern States*, Proceedings of the Conference on Wetlands of the Chesapeake, Easton, Maryland, Environmental Law Institute, pp. 143–154.

Ogawa, H and JW Male (1986) "Simulating the Flood Mitigation Role of Wetlands," *Journal of Water Resources Planning and Management*, **112** (1), 114–128.

Olshansky, RB and JD Kartez (1998) "Managing land use to build resilience," In *Cooperating with Nature: Confronting natural hazards with land use planning for sustainable communities*, ed. RJ Burby, Joseph Henry Press: Washington, DC.

Owen, CR and HM Jacobs (1992) "Wetland protection as land-use planning: the impact of section 404 in Wisconsin, USA," *Environmental Management*, **16**, 345–353.

Pasterick, ET (1998) "The national flood insurance program," In *Paying the Price: The status and role of insurance against natural disaster*, eds H Kunreuther, S Richard, and J Roth, Joseph Henry Press: Washington, DC, pp. 125–155.

Paul, MJ and JL Meyer (2001) "Streams in the urban landscape," *Annual Review of Ecological Systems*, **32**, 333–365.

Peacock, WG, BH Morrow, and H Galvin, eds (1997) *Hurricane Andrew: Ethnicity, Gender and the Sociology of Disaster*, Routledge: London, UK.

Peacock, WG, JE Kang, R Husein, GR Burns, C Prater, SD Brody and T Kennedy (2009) *An Assessment of Coastal Zone Hazard Mitigation Plans in Texas. Technical Report 09–01R*, Hazard Reduction and Recovery Center, Texas A&M University: College Station, TX.

Pielke, RA (1996) *Midwest Flood of 1993: Weather, Climate, and Societal Impacts*, National Center for Atmospheric Research: Boulder, CO.

Pielke, RA (1999) "Nine fallacies of floods," *Climatic Change*, **42** (2), 413–438.

Pielke, RA, MW Downton, and JZB Miller (2002) *Flood Damage in the United States, 1926–2000: A Reanalysis of National Weather Service Estimates*, Environmental and Societal Impacts Group, National Center for Atmospheric Research: Boulder, CO.

Platt, RH (1999) *Disasters and Democracy: The Politics of Extreme Natural Events*, Island Press: Washington, DC.

Rasmussen, PP and CA Perry (2000) *Estimation of Peak Streamflows for Unregulated Rural Streams in Kansas*, U. S. Geological Survey, Water-Resources Investigations Report 00–4079.

Rose, S and N Peters (2001) "Effects of urbanization on streamflow in the Atlanta area (Georgia, USA): A comparative hydrological approach," *Hydrological Proceedings*, **14**, 1441–1457.

Running, SW and PE Thornton (1996) "Generating daily surfaces of temperature and precipitation over complex topography," In *GIS and Environmental Modeling: Progress and Research Issues*, ed. MF Goodchild, GIS World Books: Ft. Collins, CO.

Sabatier, P (1995) "An advocacy coalition framework of policy change and the role of policy-oriented learning therein," In *Public Policy Theories, Models, and Concepts: An Anthology*, ed. D McCool, Prentice Hall: Englewood Cliffs, NJ, pp. 412.

Sacks, P (1980) "State structure and the asymmetrical society," *Comparative Politics*, **12** (3), 349–376.

Saxton, KE and SY Shiau (1990) "Surface waters of North America; influence of land and vegetation on streamflow," in *The Geology of North America Vol. 0–1, Surface Water Hydrology*, eds MG Wolman, and HC Riggs, The Geologic Society of America, pp. 55–80.

Schön, DA (1983) *The Reflective Practitioner: How Professionals Think in Action*, Basic Books: New York, NY.

Schumm, SA (1956) Evolution of drainage systems and slopes in Badlands at Perth Amboy, N.J, *Bulletin of the Geological Society of America*, **871**, 597–646.

Schuster, WD, J Bonta, H Thurston, E Warnemuende, and DR Smith (2005) "Impacts of impervious surface on watershed hydrology: a review," *Urban Water Journal*, **2** (4), 263–275.

Seaburn, GE (1969) *Effects of Urban Development on Direct Runoff to East Meadow Brook, Nassau County, Long Island, New York*, USGS Professional Paper 627-B, pp. 14, U.S. Geological Service: Washington, DC.

Sifneos, JC, EW Cake Jr, and ME Kentula (1992) "Effects of section 404 permitting on freshwater wetlands in Louisiana, Alabama, and Mississippi," *Wetlands*, **12**, 28–36.

Stein, ED and RE Ambrose (1998) "Cumulative impacts of Section 404 Clean Water Act permitting on the riparian habitat of the Santa Margarita, California watershed," *Wetlands*, **18** (3), 393–408.

Stein, J, P Moreno, D Conrad, and S Ellis (2000) *Troubled Waters: Congress, the Corps of Engineers, and Wasteful Water Projects*, Taxpayers for Common Sense and National Wildlife Federation: Washington, DC.

Steiner, F, S Pieart, E Cook, J Rich, and V Coltman (1994) "State wetlands and riparian area protection programs," *Environmental Management*, **18** (2), 183–201.

Stevens, M, Y Song, and P Berke (2009) "New Urbanist developments in flood-prone areas: safe development, or safe development paradox?," *Natural Hazards*, **53** (3), 605–629.

Stuckey, MH (2006) *Low-flow, base-flow, and mean-flow regression equations for Pennsylvania stream*, U.S. Geological Survey, Scientific Investigations Report, 2006–5130.

Szaro, R, W Sexton, and C Malone (1998) "The emergence of ecosystem management as a tool for meeting people's needs and sustaining ecosystems," *Landscape and Urban Planning*, **40** (1), 1–7.

Texas State Data Center and Office of the State Demographer (2008) Available at: http://txsdc.utsa.edu/ (accessed February 15, 2010).

Thampapillai, DJ and WF Musgrave (1985) "Flood damage and mitigation: a review of structural and non-structural measures and alternative decision frameworks," *Water Resources Research*, **21**, 411–424.

Tourbier, JT and R Westmacott (1981) *Water resources protection technology: A handbook of measures to protect water resources in land development*, The Urban Land Institute: Washington, DC.

United States Army Corps of Engineers (USACE) (2001) *2001 Annual Regulatory Statistical Data*, available at: www.usace.army.mil/CECW/Pages/reg_archives.aspx, (accessed May 6, 2008).

United States Army Corps of Engineers (USACE) (2002) "Services to the Public: Flood damage reduction", available at: www.corpsresults.us/docs/VTNFloodRiskMgmtBro_loresprd.pdf (accessed July 7, 2005).

United States Department of Agriculture (2000) *Summary Report: 1997 Natural Resources Inventory*, Natural Resources Conservation Service, USDA: Washington, DC.

United States Geological Survey (USGS) (1997) *National Water Summary on Wetland Resources*. U. S. Geological Survey, Water Supply Paper 2425.

Verry, ES and DH Boelter (1978) "Peatland hydrology," In *Wetland Functions and Values: The State of Our Understanding*, eds, PE Greeson, JR Clark, and JE Clark. American Waterworks Association: Minneapolis, pp. 389–402.

Walker, B and D Salt (2006) *Resilience Thinking: Sustaining Ecosystems and People in a Changing World*, Island Press: Washington, DC.

Weir, M and T Skocpol (1985) "State structures and the possibilities for 'Keynesian' responses to the Great Depression in Sweden, Britain and the United States," In *Bringing the State Back In*, eds PB Evans, D Rueschemeyer, and T Skocpol, Cambridge University Press: New York, NY, pp. 107–168.

Westley, F (1995) "Governing design: the management of social systems and ecosystems management," In *Barriers and Bridges to the Renewal of Ecosystems and Institutions*, eds LH Gunderson, CS Holling, and SS Light, Columbia University Press: New York, NY, pp. 391–427.

Whipple, W (1998) *Water Resources: A New Era for Coordination*, ASCE Press: Reston, VA.

White, GF (1936) "Notes on flood protection and land use planning," *Planners Journal*, **3** (3), 57–61.

White, GF (1945) "Human adjustment to floods: a geographical approach to the flood problem in the United States," In *Research Paper no. 29*, University of Chicago, Department of Geography: Chicago, IL.

White, GF, WAR Brinkmann, HC Cochrane, and NJ Ericksen (1975) *Flood Hazard in the United States: A Research Assessment*, Institute of Behavioral Sciences, University of Colorado: Boulder, CO.

White, MD and KA Greer (2006) "The effects of watershed urbanization on the stream hydrology and riparian vegetation of Los Peñasquitos reek, California," *Landscape and Urban Planning*, **74**, 125–138.

Wondolleck, JM and SL Yaffee (2000) *Making Collaboration Work: Lessons from Innovation in Natural Resource Management*, Island Press: Washington, DC.

Zahran, S, SD Brody, WG Peacock, A Vedlitz, and H Grover (2008) "Social vulnerability and the natural and built environment: a model of flood casualties in Texas, 1997–2001," *Disasters*, **32** (4), 537–560.

Index

194

Printed in the United States
by Baker & Taylor Publisher Services

Printed in the United States
By Bookmasters